Learning to Flourish in the Age of AI

This timely book affirms that humans can flourish in the Age of AI by relying on their distinctive strengths, and explores the skills and knowledge that are required to interact effectively, efficiently, and responsibly with AIs, both today and in the future.

In Part I, this book develops the "Cognitive Amplifier Loop," which allows humans to use AI to build on their cognitive and emotional strengths and manage their limitations. Kosslyn discusses ways to employ this loop to offload tasks to AI and to utilize it to train us effectively and efficiently, as well as how to use it to both learn and engage in critical thinking, creative problem solving, and manage cognitive and emotional constraints. Part II establishes how to draw on the Cognitive Amplifier Loop to help us improve our human relationships, addressing emotional intelligence, effective communication, leadership, followership, and collaboration skills. Finally, Part III builds on previous chapters to consider how to interact with AIs to help each of us learn and grow. Throughout the book, Kosslyn provides practical strategies and AI-assisted exercises to help readers develop these essential skills and knowledge. Kosslyn argues that by cultivating distinctively human capabilities, we can not only coexist with AI but flourish in an AI-infused world.

This book is a must-read for anyone involved with AI, education, or Learning and Development in industry. It will also appeal to anyone studying thinking and decision making, AI and psychology, and the use of technology in the classroom.

Stephen M. Kosslyn is the founder, CEO, and President of Active Learning Sciences, Inc., and Lindsley Professor Emeritus at Harvard University. He previously was the Founding Dean and Chief Academic Officer of Minerva University. Before that, he was the Director of the Center for Advanced Study in the Behavioral Sciences and Professor of Psychology at Stanford University, USA, after having been Chair of the Department of Psychology and Dean of Social Science at Harvard University.

Learning to Flourish in the Age of AI

Stephen M. Kosslyn

Routledge
Taylor & Francis Group

NEW YORK AND LONDON

First published 2025
by Routledge
605 Third Avenue, New York, NY 10158

and by Routledge
4 Park Square, Milton Park, Abingdon, Oxon, OX14 4RN

Routledge is an imprint of the Taylor & Francis Group, an informa business

Library of Congress Cataloging-in-Publication Data
Names: Kosslyn, Stephen Michael, 1948– author.
Title: Learning to flourish in the age of AI / Stephen M. Kosslyn.
Description: New York, NY : Routledge, 2025. |
Includes bibliographical references and index. |
Identifiers: LCCN 2024030653 (print) | LCCN 2024030654 (ebook) |
ISBN 9781032686646 (hardback) | ISBN 9781032686660 (paperback) |
ISBN 9781032686653 (ebook)
Subjects: LCSH: Artificial intelligence–Social aspects. | Critical thinking. |
Human-computer interaction.
Classification: LCC Q335 .K687 2025 (print) | LCC Q335 (ebook) |
DDC 303.48/34–dc23/eng/20240928
LC record available at https://lccn.loc.gov/2024030653
LC ebook record available at https://lccn.loc.gov/2024030654

ISBN: 978-1-032-68664-6 (hbk)
ISBN: 978-1-032-68666-0 (pbk)
ISBN: 978-1-032-68665-3 (ebk)

DOI: 10.4324/9781032686653

Typeset in Optima
by Newgen Publishing UK

Access the Support Material: www.routledge.com/9781032686660

To Dr. Robin S. Rosenberg, who makes my life possible.

Contents

PART III
Expanding Horizons in the Age of AI   241

Preface

This book is about what we humans need to learn in order to flourish in the Age of AI. My overarching claim is that we need to learn both the skills and knowledge that complement what AIs can do well and the skills and knowledge that allow us to interact with AIs effectively, efficiently, and responsibly.

I cannot possibly do justice to everything we need to learn. But I can give the reader a taste of the different topics, enough so that they can decide whether to follow up by considering the resources I cite—including many free YouTube tutorials, such as those cited at the end of each chapter. This book can help the reader to organize a comprehensive program of self-education, which will go a long way toward preparing them for what is to come. To make the prompts and tables easier to use, the reader can cut-and-paste them at: www.routledge.com/9781032686653.

My previous books on applied subjects range from how to design visual displays, to how to design slideshow presentations, to how to design active learning exercises for online courses, to how to design, deliver, and assess active learning with AI. In all cases, I focused on methods and techniques, not content. This book is different: It focuses on the specific skills and knowledge that we all need to learn in order to flourish in the Age of AI.

I use that term "flourish" intentionally. My goal here is not just to provide a survival guide, nor to provide helpful hints and best practices. Instead, I have developed a systematic approach that is generative: It should allow you to work with AIs to accomplish your own goals. I aim to help you, the reader, have a better life by using AI.

Given the theme of, and recommendations in, this book, I would be hypocritical if I eschewed using AI to help me write. And in fact, I often employed the "Cognitive Amplifier Loop" described in Chapter 2. I found AI particularly helpful for suggesting examples and analogies, for editing, and for providing overviews of topics and the names of key contributors. I did not use AI to outline or structure any part of the book, nor did I ever allow it to have the last word: In every case, I edited and revised what the AI suggested—I practiced what I preach in Chapter 2. The only exception to this is when I provide transcripts of interactions with an AI, which are verbatim records of what actually happened.

That said, the contributions of AI to this book are dwarfed by the contributions of many friends, colleagues, and family members: As always, my wife Dr. Robin Rosenberg offered sage advice, insightful critiques, and detailed editorial suggestions. My children Justin, David, and Neil all took time to argue with me and set me straight on technical points—all three of them have jobs that require deep mastery of AI, and I benefited from their expertise in more ways than they probably realize. I also thank Dr. Anne Marie Ward and Heyu Huang, who gave me insightful comments on early drafts, which helped shape the final project. I also need, once again, to thank Dr. Beth Callaghan, my friend and colleague at Active Learning Sciences, who read a draft and effectively played devil's advocate. Her constructive and creative criticism made this book much better. And Dr. Kacey Warren provided a very astute and detailed critique of an earlier version, and helped me focus and be clearer. She brought to bear her formidable "philosopher's perspective" and made me think hard about what I was saying—which, among other things, led me to delete an entire chapter. I must note, however, that any errors or glitches are entirely my responsibility. I also thank Niki Gallagher Garcia for research assistance and proofreading. And, of course, I thank Tori Sharpe and Alison Macfarlane at Routledge and copyeditor Tom Bedford, who were patient and helpful at every turn.

Chapter 1

Being Human in the Age of AI

OpenAI, a Silicon Valley startup, released its ground-breaking ChatGPT software on November 30, 2022. This new type of artificial intelligence (AI) immediately led to a mixture of shock, awe, excitement, and—I think it's fair to say—fear and trepidation. Within two months, over 100 million people became active users, which was a record-setting speed of adoption of a new technology.[1] People enthusiastically embraced this new type of AI for good reason: It writes remarkably well on a wide range of topics, and its descendants—from the likes of Anthropic, Google, Meta, and OpenAI— can do everything from creating a plausible syllabus for a business school course, to writing a passable sermon, to suggesting clever titles for a movie based on a few bullet points about its plot, to analyzing job descriptions to infer prerequisite competencies, to writing poetry, to synthesizing ideas, to writing publishable essays on a wide range of topics. Moreover, these AIs take direction well: If we don't like the initial draft, we can ask them to improve the product. For example, we can ask them to make the tone more formal, provide more examples, focus more on a particular aspect, and/or change the level of sophistication. Even the initial version of ChatGPT did all of this remarkably well.[2]

In general, these new AIs are good at analyzing, evaluating, synthe- sizing, and creating new things by building on previous knowledge. In fact, we can give them a grading rubric and ask them to use that rubric to grade learner essays, and they assign grades similarly to how a human instructor would assign them.[3] The new AIs can also interpret and produce images. We can upload a photograph of our open refrigerator door and ask the AI to suggest recipes that draw on what we have, and it accommodates. We can ask them to show us pictures of different sorts of cats, and out they come. Indeed, we can even ask for an image of a non-existent animal that is part cat and part dog, and it easily creates the hybrid. Barely a week goes by without an announcement of some new amazing thing that AIs can do,

DOI: 10.4324/9781032686653-1

from detecting cancer and predicting dementia to being able to convert descriptions to fancy infographics or lifelike videos.[4]

The new AIs are examples of *generative artificial intelligence*. They produce something new, based on massive amounts of training on relevant data. In this book we focus on such generative AIs, which from here on I will simply refer to as "AIs." The progress in this field has been stunning since the release of ChatGPT. The AI companies introduce innovations so quickly that many people feel they don't have time to catch their breath. Indeed, less than six months after the release of ChatGPT, over 1,000 researchers in AI and practitioners in related fields signed a petition asking that all research on larger, more capable AI models be paused for six months to give researchers—and society at large—a chance to reflect on how we will control such technology.[5] People were worried about everything from AIs eliminating jobs to their going into Full Terminator Mode and taking over the world and exterminating humans. Given the cutthroat competition among the entrant AI companies, this petition was doomed from the start, but it did reflect just how ground-breaking and potentially earth-shattering the new AIs are—and how nervous many people have become about them. Following up, various governments have now put in place regulations that govern the development and use of AIs,[6] and more such regulations are sure to follow.

In light of such rapid progress, one danger of writing a book on AI is that it will soon be outdated. That is why I focus not on the AIs per se, but rather on their human users. This book is about what humans need to learn in order to flourish in an AI-infused world.

The Nature of the Beast

In order to flourish in an AI-infused world, we need to have a basic knowledge of how generative AIs work. Such knowledge can allow us to harness their strengths and mitigate their limitations. Thus, we begin with a quick overview of essential characteristics of these new AIs.

Many characteristics of AIs result from the design of the software systems themselves. To see this, let's start by contrasting the new AIs with the traditional computer programs we all know, such as Adobe Acrobat, Apple Mail, Microsoft Excel, and Microsoft Word. Imagine a sheet of paper and a pencil. The sheet of paper stores representations, such as words and drawings; the pencil creates and modifies such representations according to particular rules (e.g., grammar for text and foreshortening for images). Traditional computer programs are similar: They store data in specific places in a computer's memory, and then operate on that data according to specific rules. For example, a standard spreadsheet program (such as

Microsoft Excel or Google Sheets) operates on stored digits and can apply the same set of rules, such as addition and multiplication, to any set of digits. If it stores 20212 and 1640, it can add or multiply them, and the same goes for 175 and 11, and so forth. It doesn't matter what the digits are; as long as they are digits, the machine can apply the rules of arithmetic to them. And the program applies these rules reliably with great precision, getting the same result for the same operation on the same digits, no matter how many times it does this.

In contrast, the new AIs are a different type of computer system, which does not have a separate database of symbolic representations, such as of digits and words, that they create and process according to stored rules, such as those of arithmetic and grammar. Rather, these AIs are based on artificial neural networks, which are loosely modeled on how the brain works. These artificial networks include a set of units, "artificial neurons," often millions of them. These artificial neurons are arranged into layers, and each unit is connected to many other units in the adjacent layers. These connections can be positive or negative. If one unit has a positive connection to another, activating the first unit leads it to try to activate the second unit. In contrast, if one unit has a negative connection to another, activating the first unit leads it to try to prevent the second one from being activated. There are many, many, many such connections among the artificial neurons. Moreover, the connections vary in their strengths: The greater the strength, the more forcefully one unit either activates or inhibits the other, depending on whether the connection is positive or negative.

In neural networks, any particular piece of information corresponds to a specific pattern of strengths of connections among many units. Information is not localized the way it is in a traditional computer program. Rather, information is distributed across many connections. Indeed, any given connection may be part of multiple patterns that store different pieces of information—somewhat like the way a given letter can be part of more than one word in a crossword puzzle. Moreover, neural networks don't operate the way standard computer programs do. The neural network as a whole acts as a kind of giant filter: A set of "input units" is stimulated from the outside (e.g., as occurs when we ask an AI a question), which activates or inhibits the connected units, in accordance with the nature of the connections. The wave of activation flows forward, and sometimes laps backward, percolating through the network and ultimately producing a specific output. The particular output that results from a specific input depends on the nature and the strengths of the connections among the units.

What determines the strengths of the connections? In a word, training. When an AI is being trained to process language, the network is initially given massive amounts of text and it tries to predict successive "tokens," which are the smallest unit of text that the network processes (such as short words, parts of words, or symbols). The model relies on the preceding tokens to predict the next one. This prediction is done automatically, and because the network is fed the entire words and sentences, it can check whether it guessed correctly and adjust the strengths among the relevant connections to correct errors. Once the network gets very good at such anticipation, the networks are trained on more specific materials to "fine tune" them. Following this, humans step in for a final round of training. These human trainers engage in "Reinforcement Learning from Human Feedback" to encourage certain outputs and discourage others. If the human trainers like a response, they cause the network to strengthen the connections that led to it; if they don't like a response, they cause the network to weaken positive connections and strengthen negative ones. This is where "guardrails" come from, which are supposed to prevent the network from providing dangerous information or advice, such as helping users to build bombs or commit crimes.

These new AIs generally have a "GPT architecture." The "G" stands for "Generative"; the model does not simply look up previously stored information, the way Google does or the "Find" function in a word processing program does, but rather relies on patterns of connections to generate novel responses. The "P" stands for "Pre-trained"; the model was trained in advance, often on virtually everything on the internet, usually supplemented by other sources. And the "T" stands for "Transformer"; for language, the network interprets the tokens and pays attention to where two (or more) tokens occur in a sentence, transforming the input into specifications about the meanings of specific tokens and their relationships in the sentence. This transformation process allows the network to give the illusion of "making sense" of the input.

This is a very quick overview of a complex topic, but should suffice for present purposes. Many highly knowledgeable researchers have provided detailed tutorials on neural networks and GPT architecture (e.g.,[7,8,9]).

Although generative AI systems that create and interpret images rely on the same underlying principles as the Large Language Models (LLMs) we have discussed so far, we focus on LLMs because these AIs are most likely to fundamentally change how we live and work.

Living and Working in an AI-Infused World

We can already use the internet to access much—if not most—of the world's knowledge, accumulated over all of human history. For most purposes, the "cloud" can serve as an extended memory store for our brains;

it retains information so we won't need to do so ourselves. Moreover, AIs are already at least as good as most humans at various cognitive skills that require processing information, such as critical thinking, creative thinking, solving well-defined problems, making well-structured decisions, recognizing patterns, and making many sorts of predictions. AIs can already carry out many cognitive tasks, and they are only going to get better as new models are released. AIs can process a lot of information, so we won't need to do so ourselves.

These observations lead to profound questions: If we can find on the cloud whatever information we need when we need it, and AIs can process that information as we desire, why should we bother to acquire such knowledge and skills ourselves? What should we learn instead? What do we need to know to flourish in this emerging world?

Here is a simple observation that strikes to the heart of these issues: *Humans are good at responding in open-ended situations that require taking context into account* (cf. [10,11,12]). In "open-ended situations," we don't know in advance which factors will play a central role, defining the situation. In fact, new factors can suddenly arise, such as a fire alarm going off or a demonstration in the street, which can turn a routine visit to the post office into an ordeal. And "context" refers to the factors that frame or surround a given situation, which affect how we interpret it. We hear a loud thump differently in the dead of night than during a bright afternoon, and we might laugh if someone has a slip of the tongue in a casual conversation but would cringe if they said the same thing during a eulogy. We humans can turn on a dime to adapt as the situation changes in unanticipated ways, and we are extremely good at modulating what we perceive, think, feel, and how we behave depending on the context. Moreover, we can learn to be even better at these skills (e.g., [13]).

Although AIs are very effective in well-defined situations, such as when they provide advice about interpreting scores on specific tests, they falter when the situation is open-ended and context must be taken into account.[14,15,16] In the words of Harvard Professor Gary King (personal communication), AIs are good at interpolating, but they are bad at extrapolating.

The struggles AIs have with such situations are not simply a result of their not being updated often enough. Rather, the way we humans interpret situations and events relies not just on our past experiences, but also on the way our human brains and bodies function. Indeed, we rely on specific facets of our brains and bodies that can give rise to hunches and intuitions, which would require an AI to extrapolate beyond its training set.

To make this concrete, let's consider classic studies by Antonio Damasio's research group.[17,18,19,20,21,22,23] They devised a game that relies on four card decks, and each card indicates a monetary gain or monetary loss. The goal is to maximize gains by figuring out which decks are most likely to lead to gains vs. losses. The cards in two of the decks, A and B, produce high immediate gains but then include cards that lead to very high losses and produce net losses over time. In contrast, the cards in the other two decks, C and D, produce smaller gains but also less frequent losses, which results in overall gains over time.

After a few rounds of drawing cards, the participants preferred to draw cards from Decks A and B, which led to initial gains but then accumulated steady losses. After 20–40 turns, these people switched to Decks C and D. What was most interesting is what happened after they were deep into the game and were about to reach for Deck A or B, which were likely to incur a loss: In this case, the participants had physiological responses—such as increased sweat in their palms—that signal anxiety. The players apparently had a "hunch" that something was amiss with Decks A and B, and they had this sense well before they consciously inferred what was going on.

Such hunches appear to arise from a specific part of the brain, the ventromedial prefrontal cortex (which is in the front, middle, lower part of the brain). Indeed, patients who had suffered damage to this brain structure never showed anticipatory emotional responses and never learned to shift to the better decks. Based on their findings with brain-damaged patients, the researchers inferred that the ventromedial prefrontal cortex plays a crucial role in this unconscious decision process. This brain structure appears to cross-index emotional reactions and the results of logical thinking, which can give rise to hunches and intuitions that nudge us to behave in certain ways well before we can explain why we make such decisions.

AIs, not having emotions or bodies, cannot make use of this sort of mechanism. These hunches and intuitions can occur in open-ended situations where context is important—such as novel made-up card games—and thus AIs cannot be trained on all of them in advance.

Moreover, their lack of bodies, hormones and the like implies that current AIs have another set of limitations. Consider an insight by the Viennese philosopher Ludwig Wittgenstein,[24] who commented that "Even if a lion could speak, we could not understand him." This disconnect arises because lions have a different "form of life" than humans. For example, they run on four legs instead of standing upright on two legs, have long sharp claws instead of fingers, navigate through tall grass, and so on. These differences lead them to conceptually "carve up" the world differently than humans do, and thus even if they had words, their underlying concepts would not match human concepts.

We can make the same argument regarding AIs: They don't have bodies, don't have hormones, don't experience the effects of fatigue, and so on.

Hence, they cannot fully understand what it's like to be a human. They can simulate such knowledge based on their training sets, and interpolate within that set, but cannot easily extrapolate to open-ended situations that require taking context into account. Moreover, even though we humans are all members of the same species, we are not identical—and these differences lead each of us to see things at least a little differently than other people. Thus, AIs will not carve up the world exactly as any given human, and they will stumble when trying to understand each of us in open-ended situations that require taking context into account.

I asked GPT-4 to provide examples of open-ended situations where context matters. Here is what it produced:

- *Negotiating deals and contracts.* In negotiations, being able to quickly read the room, understand the other party's reactions, and adapt our strategy accordingly can lead to more favorable outcomes. Immediate context like body language, tone, and choice of words can provide crucial insights.
- *Delivering feedback to a coworker.* When giving feedback, it's important to be attuned to the co-worker's mood, the setting, and their reactions as the conversation unfolds.
- *Managing classroom dynamics as a teacher.* Teachers must adjust their teaching methods and interventions based on student engagement, understanding, and behavior.
- *Performing theater or music.* Performers must be attuned to their audience's reactions and the overall atmosphere. Quick adjustments, such as changing the setlist or interacting with the audience, can significantly enhance the performance.
- *Conducting diplomatic engagements.* Diplomats must be acutely aware of cultural nuances, the political climate, and subtle cues in conversations.
- *Providing customer service.* Customer service professionals must adapt their approach based on the customer's immediate reactions and emotions.
- *Managing emergencies and crises.* In emergencies, such as natural disasters, accidents, or critical medical situations, quick, context-aware decisions can mean the difference between life and death.
- *Conducting combat and military operations.* Soldiers and commanders must constantly be aware of their immediate surroundings and adjust quickly to changing conditions on the battlefield.

- *Performing sports and athletics.* Athletes and coaches must be keenly aware of the immediate context during competition, including the actions of opponents, environmental conditions, and their own physical state.
- *Performing surgery.* Surgeons need to be highly responsive to the immediate context within an operation. Changes in a patient's condition, unexpected complications, and varying responses to treatment require quick and adaptive decision-making.
- *Enforcing laws.* Law enforcement officers responding to incidents must quickly assess and adapt to rapidly changing scenarios, balancing the need for immediate action with considerations for safety and legality.
- *Interpreting news reports.* A key part of deciding whether a news report is accurate is to assess how well the purported event fits into the surrounding context.

Unaided AIs are not likely to perform these sorts of tasks very well. Thus, it makes sense for us humans to learn the skills and knowledge that are required to perform them—including how to leverage AIs to help us do so. Note, however, that our human cognitive and emotional limitations—including biases—can get in the way of our doing well any of the tasks noted above. Hence, we not only need to learn specific skills and knowledge, but we also need to recognize and manage our limitations, as discussed later in this book.

Which particular skills and knowledge are relevant to carrying out such tasks? I next asked GPT-4 to indicate the underlying skills and knowledge that cut through the various tasks summarized above. I submitted the above list in a new chat session, and it replied as follows:

The tasks you've listed, though diverse in nature, draw upon a core set of underlying skills and knowledge areas. These include:

- Emotional Intelligence: Understanding and responding to others' emotions is crucial in tasks like delivering feedback, negotiating, providing customer service, and managing classroom dynamics. It involves reading non-verbal cues like body language and tone, and adjusting one's approach accordingly.

- Situational Awareness: This involves being cognizant of the immediate environment and its dynamics, a skill essential in emergency response, combat operations, sports, and live performance. It requires the ability to rapidly assess and understand the situation around oneself.
- Adaptability and Flexibility: The capacity to change strategies or tactics in response to new information or changing circumstances is key across these scenarios, from diplomatic engagements to medical surgery. This includes adapting communication styles, decision-making approaches, or physical actions.
- Critical Thinking and Analytical Skills: These skills are necessary for interpreting news articles, assessing the validity of information, and making informed decisions during crisis management or when responding to rapidly evolving situations in law enforcement.
- Cultural Competence and Contextual Knowledge: Understanding cultural nuances and the specific context of an interaction is vital in diplomacy, teaching, and customer service. This encompasses knowledge of different cultural norms, practices, and sensitivities.
- Decision Making Under Pressure: Many of these tasks, such as in emergency response, medical surgery, or combat, require making quick, decisive actions under stressful conditions. This involves weighing options rapidly and effectively, often with limited information.
- Effective Communication: The ability to convey information clearly and persuasively is essential in negotiation, feedback delivery, and diplomatic engagements. This also includes the ability to listen actively and interpret the responses of others.
- Psychological Understanding: A good grasp of human behavior and psychology aids in understanding reactions in various contexts, from classroom dynamics to customer interactions. This knowledge helps in tailoring approaches to individual or group needs.
- Physical Coordination and Awareness: Particularly in sports, live performances, and combat operations, physical awareness and coordination are critical. This involves understanding and controlling one's own body movements and being attuned to physical surroundings.
- Professional Expertise: Specific knowledge relevant to the task at hand, such as medical knowledge for surgeons or legal knowledge for law enforcement, is crucial. This expertise forms the foundation upon which the other skills are applied.

As I argue in the following chapter, we should not treat such responses from an AI as gospel, but rather as a place to start. In this case, if we utilize the simple criterion of seeking situations that are open-ended and require taking context into account, we get to much the same place as what the AI suggests. AIs are not likely to supplant us humans at such skills and knowledge in the near future, given that they rest on capabilities that humans can excel at and contemporary AIs do not.

Flourishing in the Age of AI

What, exactly, are we aiming for in this book? What would it mean to "flourish" in the Age of AI? Much of what we need to learn was already important prior to the rise of AI, but will become increasingly important as other people come to rely on AIs and the culture as a whole incorporates them into daily life.

Various thinkers have noted that flourishing requires us to achieve numerous goals (e.g.,[25,26,27,28,29]). These goals include achieving a sense of autonomy and control, having fulfilling relationships, feeling satisfied at work, having enough money, and having a good work–life balance. In addition, flourishing requires personal growth and developing our abilities and talents. And above and beyond just feeling satisfied, we need life goals that give us a sense of purpose and meaning.

How do AIs figure into achieving these goals? We need to consider two kinds of factors: On the one hand, we must identify ways that AIs can help us achieve these ends. On the other hand, we need to identify ways to avoid having AIs undermine these goals.

Let's consider each goal in turn. In what follows, we use only the broadest strokes, to set the stage for what is to come in later pages.

Sense of Autonomy and Control

To flourish, we need to have a sense of autonomy and control. One crucial aspect of achieving this, going forward, is building skills and knowledge that will allow us to interact effectively with AIs. We need to know how and when to initiate such interactions, how to interpret the AI's claims, products, and suggestions, and how to steer AIs in directions we desire. Clearly, becoming strong critical thinkers and creative problem solvers is key. In addition, we need to know how to manage our cognitive and emotional limitations, such as those imposed by various biases.

We also need to know how to use these skills to deal with problems and issues that AIs will present, so that we don't feel helpless or buffeted by their effects. For example, major tech companies, such as Meta and Google, are grappling with ways to regulate disinformation and misinformation. Unfortunately, this is an arms race that cannot be won: As companies develop algorithms to spot such misleading information, the perpetrators get better at disguising their works. Indeed, "deep fakes" are getting so good that many experts think it will soon be nearly impossible to identify them.[30,31,32] The only solution to these problems is to recognize that trying to completely regulate the source is like putting our finger in a dike and hoping that we won't have to stand there the rest of our lives—and hoping that the hole stays at a convenient finger-plugging size and that not too many other holes pop up. Some such leak-plugging is probably necessary, but it's also important to look at the other side of this equation: Instead of focusing solely on the source, let's also consider the receiving end. Rather than just trying to plug holes in the dike, we should also learn to swim.

Fulfilling Relationships

Perhaps paradoxically, in the Age of AI we will probably place an increasingly large premium on our personal relationships. We will delegate much to AIs, and what is left over will be increasingly important to us. Thus, more than ever, we need the skills and knowledge to interact effectively with other humans. This requires developing our leadership, followership, collaboration, and communication skills. To do so effectively, we need to infer the cognitions, emotions, and motivations of others. We must grasp the ways that personality traits vary, how they influence behavior, and the ways that culture affects us. Based on all of this and more, we must develop and exercise emotional intelligence.

We also need to be alert to temptations that may undermine these goals. In particular, we need to avoid becoming AI addicts, who spend all of our time glued to our screens. We must appreciate what interacting with other people can do for us, which requires appreciating what gives meaning and purpose to our lives.

Feeling Satisfied at Work

AIs will probably eliminate or transform many different forms of work, running the gamut from manual to intellectual jobs.[33,34] AIs will soon power mobile robots, which may displace gardeners, workers in warehouses, kitchen help in restaurants, and even perhaps soldiers in the military. AIs will also displace members of the professions, such as medicine and the law.

For example, AIs already can diagnose many illnesses better than human physicians,[35,36] and AIs can write contracts, analyze legal cases, and carry out many functions of paralegals and attorneys.[37,38] Such developments are likely to accelerate.

To remain relevant in a world suffused with AIs, we need to capitalize on our human ability to function in open-ended situations that require taking context into account. We may need to shift from focusing on the *quantity* of work performed to focusing on the *quality* of what we do. We can gain considerable satisfaction at work if we know how to revise and refine the products and suggestions of AIs. We can also gain satisfaction at work by providing high-quality judgments and decisions, particularly in the sorts of open-ended situations that are challenging for AIs.

Moreover, we need to do this as the world rapidly changes. To achieve this goal, we must know how to recognize when we need to learn and must be able to acquire skills and knowledge effectively and efficiently as jobs are transformed or replaced. Indeed, some have claimed that even now—before AIs have deeply penetrated the workplace—"The average half-life of skills is now less than five years, and in some tech fields it's as low as two and a half years."[39] Learning should become a way of life, which will be necessary to meet the ever-changing moment.

Work–Life Balance

AIs are going to perform many jobs that humans now do, especially those jobs that are highly structured. To put a positive spin on this, humans will have more free time. At least for some people, the additional free time may prove challenging: We need to discover what is important to us as individuals, which requires understanding how to set personal life goals and work toward them.

Having Enough Money

If we put in fewer hours at our jobs, does this imply that we will make less money? Not necessarily. As AIs automate increasingly more work, the high-quality work we humans can contribute may become increasingly valuable—and hence paid better. In addition, if we learn to deploy AIs effectively and efficiently, we can make more money from the same amount of effort.

On the flip side, one thing is increasingly clear: If we don't master the skills and knowledge needed to interact effectively with AIs, and if we don't hone the human abilities that will be increasingly valued, we are not likely to make more money.

Personal Growth and Developing Our Abilities and Talents

AIs can also help us grow as people. AIs can serve, within limits, as personal coaches and tutors, and they can personalize many types of instruction.[40] We can take advantage of this to help us develop our abilities and talents.

However, to take full advantage of our human cognitive strengths, we need to learn to recognize and manage our cognitive limitations and our emotions. As we shall see in later chapters, AIs can help us to do so. We also need to note our biases and the factors that underlie our motivations, and approach them with a critical eye. AIs can individualize instruction that helps us grow in these ways.

Purpose and Meaning in Life

Finally, we must formulate cogent personal life goals and approach them strategically. To make such goals a true North Star, giving our lives purpose and meaning, we need to make decisions that are ethically coherent and aligned with our personal values. Formulating clear and coherent life goals is especially important in the Age of AI because although AIs can process information, humans are responsible for ensuring that actions are value-driven. Everything in this book builds to helping us reflect on such goals, which we consider in the final chapter.

In addition to the factors we just considered, we need to acknowledge that broader social-contextual conditions also affect whether we can flourish. Although these are important considerations, we have a narrower focus here. In this book we concentrate on factors that are—at least to a large extent—directly under our control. We want to identify what we can learn that will help us flourish in the Age of AI.

The following chapters are organized into three parts. Part I consists of Chapters 2, 3, 4, and 5. In Chapter 2, we develop a method for interacting with AIs, which is then extended in the following chapters. This method goes a long way toward addressing our need for autonomy and control, and can help us perform better at our jobs. We rely on this method throughout the remainder of this book. Part II consists of Chapters 6, 7, 8, and 9. In this part of the book we consider how to use our method for interacting with AIs to help us improve our human relationships. We see that the same capacities needed to interact effectively with AIs serve us well when we need to navigate an AI-infused social world. We can use AIs to help us interact more effectively with other humans, who themselves are charting their paths through this world. Finally, Part III consists of Chapters 10 and 11. We build on the material in the earlier chapters to help us adapt and grow as the world changes, not only learning but also defining and adjusting our life goals.

Digging Deeper

For readers who want to dig deeper into the subject matter, they can of course consult the articles, books, and videos that are specifically cited. However, much of this material is technical and intended for specialists. Thus, each chapter ends with a "Digging Deeper" section. This section contains examples of accessible videos that provide a more in-depth treatment of aspects of the material reviewed in the chapter. In addition, because videos come and go on the internet, and the reader might want to dig deeper still, I've also included search terms that can help the reader locate new videos on the topic.

Generative AI explained in 2 minutes: https://www.youtube.com/watch?v=rwF-X5STYks

How will AI change the world? https://www.youtube.com/watch?v=RzkD_rTEBYs

- "Generative AI explained"
- "ChatGPT impact on society"
- "AI language models transforming industries"
- "Ethical concerns with advanced AI"
- "Government regulations on AI development"

How Deep Neural Networks Work – Full Course for Beginners: https://www.youtube.com/watch?v=dPWYUELwIdM

Neural Network in 5 Minutes | What Is a Neural Network? | How Neural Networks Work | Simplilearn: https://www.youtube.com/watch?v=bfmFfD2RIcg

- "Differences between traditional programming and AI"
- "How artificial neural networks work"
- "Artificial neurons and connections in neural networks"
- "Symbolic AI vs. neural networks"
- "Strength of connections in neural networks"

How AI Could Empower Any Business | Andrew Ng | TED: https://www.youtube.com/watch?v=reUZRyXxUs4

How to Leverage AI for Personal Growth: https://www.youtube.com/watch?v=S4kSR-pB83c

The social dilemma: www.netflix.com/title/81254224

- "Developing critical thinking skills in the Age of AI"
- "Importance of human relationships in the Age of AI"
- "Job satisfaction and the changing nature of work with AI"
- "Achieving work–life balance in the Age of AI"
- "Personal growth and development with AI assistance"
- "Finding meaning and purpose in the Age of AI"
- "Ethical decision making and AI"
- "Developing emotional intelligence in the Age of AI"
- "Continuous learning and adapting to AI-driven changes"
- "Leveraging AI for personal growth and talent development"

The limits of AI and ChatGPT: The common sense problem: https://www.youtube.com/watch?v=w9MSsnoTSJs

- "Advantages of human intelligence over artificial intelligence"
- "AI limitations in understanding context"
- "Human skills that AI cannot replace"
- "Situational awareness and adaptability in humans vs. AI"
- "Decision making under pressure: humans vs. AI"
- "Cultural competence and AI limitations"
- "Effective communication skills in the Age of AI"

References

1   Hu, K. (2023, February 2). ChatGPT sets record for fastest-growing user base—analysis note. *Reuters*. www.reuters.com/technology/chatgpt-sets-record-fastest-growing-user-base-analyst-note-2023-02-01/

2   Zsolnai-Feher, K. (2023). *OpenAI ChatGPT: The future is here!* https://www.youtube.com/watch?v=V2RoqUr0qDU&t=1s

3   Kosslyn, S. M. (2023). *Active learning with AI: A practical guide*. Alinea Learning.

4   Mollick, E. (2024). One useful thing. *Substack*. www.oneusefulthing.org/

5   Associated Press. (2023, March 29). Tech leaders urge a pause in the "out-of-control" artificial intelligence race. *NPR*. www.npr.org/2023/03/29/1166896809/tech-leaders-urge-a-pause-in-the-out-of-control-artificial-intelligence-race

6   Shein, E. (2024, March 15). Governments setting limits on AI. *Communications of the ACM*. https://cacm.acm.org/news/governments-setting-limits-on-ai/

7   Karpathy, A. (2023). *Let's build GPT: From scratch, in code, spelled out*. www.youtube.com/watch?v=kCc8FmEb1nY

8   Twarog, A. (2022). *ChatGPT tutorial—A crash course on ChatGPT for beginners*. www.youtube.com/watch?v=JTxsNm9IdYU

9   Wolfram, S. (2023, February 14). *What is ChatGPT doing… and why does it work?* Stephen Wolfram Writings. https://writings.stephenwolfram.com/2023/02/what-is-chatgpt-doing-and-why-does-it-work/

10   Evans, J. S. B. T. (2002). Logic and human reasoning: An assessment of the deduction paradigm. *Psychological Bulletin, 128,* 978–996.

11   Holyoak, K. J., & Cheng, P. W. (2011). Causal learning and inference as a rational process: The new synthesis. *Annual Review of Psychology, 62,* 135–163.

12   Sarathy, V. (2018). Real world problem-solving. *Frontiers in Human Neuroscience, 12,* 261. doi:10.3389/fnhum.2018.00261.

13   Acar, O. A. (2023, June 6). AI prompt engineering isn't the future. *Harvard Business Review.* https://hbr.org/2023/06/ai-prompt-engineering-isnt-the-future

14   Aboze, B. J. (2024, March 25). How to measure LLM performance. *deepchecks.* https://deepchecks.com/how-to-measure-llm-performance/

15   Wei, F., Chen, X., & Luo, L. (2004). Rethinking generative large language model evaluation for semantic comprehension. https://ar5iv.labs.arxiv.org/html/2403.07872v1

16   Zaphir, L., & Lodge, J. M. (2023, June 27). Is critical thinking the answer to generative AI? *Times Higher Education.* www.timeshighereducation.com/campus/critical-thinking-answer-generative-ai

17   Bechara, A., & Damasio, H. (2002). Decision-making and addiction (Part 1): Impaired activation of somatic state in substance dependent individuals when pondering decisions with negative future consequences. *Neuropsychologia, 40,* 1675–1689.

18   Bechara, A., Damasio, A. R., Damasio, H., & Anderson, S. W. (1994). Insensitivity to future consequences following damage to human prefrontal cortex. *Cognition, 50,* 7–15.

19   Bechara, A., Damasio, H., & Damasio, A. R. (2000). Emotion, decision making and the orbitofrontal cortex. *Cerebral Cortex, 10,* 295–307.

20   Bechara, A., Damasio, H., Tranel, D., & Damasio, A. R. (1997). Deciding advantageously before knowing the advantageous strategy. *Science, 275,* 1293–1295.

21   Bechara, A., Tranel, D., & Damasio, H. (2000). Characterization of the decision-making deficit of patients with ventromedial prefrontal cortex lesions. *Brain, 123,* 2189–2202.

22   Bechara, A., Tranel, D., Damasio, H., & Damasio, A. R. (1996). Failure to respond autonomically to anticipated future outcomes following damage to prefrontal cortex. *Cerebral Cortex, 6,* 215–225.

23   Damasio, A. R. (1994). *Descartes' error: Emotion, reason, and the human brain.* G.P. Putnam.

24   Wittgenstein, L. (1953). *Philosophical investigations* (G. E. M. Anscombe, Trans.). Basil Blackwell.

25   Ben-Shahar, T. (2009). *The pursuit of perfect: How to stop chasing perfection and start living a richer, happier life.* McGraw-Hill.

26   Nussbaum, M. C. (2011). *Creating capabilities: The human development approach.* Harvard University Press.

27   Sen, A. (1999). *Development as freedom*. Oxford University Press.
28   Layard, R. (2011). *Happiness: Lessons from a new science* (2nd ed.). Penguin Books.
29   VanderWeele, T. J. (2020). Activities for flourishing: An evidence-based guide. *Journal of Positive Psychology and Wellbeing, 4*, 79–91.
30   Miller, E. J., Steward, B. A., Witkower, Z., Sutherland, C. A. M., Krumhuber, E. G., & Dawel, A. (2023). AI hyperrealism: Why AI faces are perceived as more real than human ones. *Psychological Science, 34*, 1390–1403. https://doi.org/10.1177/09567976231207095
31   Verdoliva, L. (2020). Media forensics and deepfakes: An overview. *IEEE Journal of Selected Topics in Signal Processing, 14*, 910–932.
32   Westerlund, M. (2019). The emergence of deepfake technology: A review. *Technology Innovation Management Review, 9*, 39–52.
33   Eloundou, T., Manning, S., Mishkin, P., & Rock, D. (2023). *GPTs are GPTs: An early look at the labor market impact potential of large language models*. Cornell University: arXiv:2303.10130 [econ.GN].
34   Georgieva, K. (2024, January 14). AI will transform the global economy. Let's make sure it benefits humanity. IMF blog. www.imf.org/en/Blogs/Articles/2024/01/14/ai-will-transform-the-global-economy-lets-make-sure-it-benefits-humanity
35   Kita, J. (2024, April 12). Are you ready for AI to be a better doctor than you? *Medscape*. www.medscape.com/viewarticle/are-you-ready-ai-be-better-doctor-than-you-2024a100070q?form=fpf
36   Tu, T., Palepu, A., Schaekermann, M., Saab, K., Freyberg, J., Tanno, R., et al. (2024). *Towards conversational diagnostic AI*. Cornell University: arXiv:2401.05654 [cs.AI].
37   Barton, R. E. (2023, August 23). How will leveraging AI change the future of legal services? *Reuters*. www.reuters.com/legal/legalindustry/how-will-leveraging-ai-change-future-legal-services-2023-08-23/
38   Hatzius, J., Briggs, J., Kodnani, D., & Pierdomenico, G. (2023, March 26). *The potentially large effects of artificial intelligence on economic growth*. Goldman Sachs, Global Economics Analyst. www.ansa.it/documents/1680080409454_ert.pdf
39   Tamayo, J., Doumi, L., Goel, S., Kovacs-Ondrejkovic, O., & Sadun, R. (2023, September–October). Reskilling in the Age of AI. *Harvard Business Review*. https://hbr.org/2023/09/reskilling-in-the-age-of-ai
40   Kosslyn, S. M. (2023). *Active learning with AI: A practical guide*. Alinea Learning.

Part I

Interacting with AIs

Chapter 2

Exploiting the Cognitive Amplifier Loop

In the future, many—if not most—of us will consult and interact with AIs in the normal course of our lives. Should we conceive of AIs as partners, collaborators, tools, or co-pilots? In my view, none of these characterizations captures the essence of what AIs are: Partners share responsibility, which requires intentions that AIs don't have; collaborators have volition and agency, which AIs do not currently have; and tools are usually passive and respond only as explicitly directed—which might characterize small, very limited AIs, but fails to characterize the large, very powerful Large Language Models (LLMs) of interest here. Similarly, regarding AIs as "co-pilots" is problematic, given that they are not true collaborators who can take initiative. Instead, in my view AIs are best thought of as *cognitive amplifiers*. Cognitive amplifiers are technologies that work with people *to augment their mental strengths and compensate for their mental limitations*. For example, among our mental strengths are our abilities to detect patterns and create stories, whereas among our mental limitations are our limited ability to pay attention to several things at once and our many biases, which cloud our reasoning. AIs can help us leverage our strengths and mitigate our limitations.

We can gain insight into the possible roles of AIs by considering an earlier, much more limited type of cognitive amplifier, electronic calculators. When calculators first appeared, some people viewed them with suspicion, thinking that they encouraged cheating—which has, of course, been echoed in reactions to using ChatGPT in school.[1] But in fact, electronic calculators allowed learners to concentrate more on the concepts behind math and less on rote computations. For example, calculators allowed learners to focus on what a mathematical operation—such as taking a square root—does, why it's useful, and when it's appropriate to employ. Moreover, calculators encouraged learners to acquire an "order of magnitude" sense of what the output should be. For instance,

DOI: 10.4324/9781032686653-3

if the result should be somewhere between 50 and 150 and it comes out as 1,500, they should immediately know that something is wrong. This focus on underlying concepts requires a deeper understanding of math than is required simply to memorize and carry out a set of steps. If learners are tested on their ability to draw on the concepts behind math, and know that these tests are coming, they will work to master this material. This is true even if they rely on a calculator to perform the actual computations.

AIs can function as the cognitive parallel of a calculator, but instead of needing to understand mathematical functions, the user needs to master a host of mental skills, such as knowing how to evaluate claims, how to identify inconsistencies, and how to analyze and structure an argument.

The power of an AI as a cognitive amplifier should not be underestimated. For example, consider this quote from the MIT economist David Autor:[2]

> The unique opportunity that AI offers to the labor market is to extend the relevance, reach, and value of human expertise. Because of AI's capacity to weave information and rules with acquired experience to support decision-making, it can be applied to enable a larger set of workers possessing complementary knowledge to perform some of the higher-stakes decision-making tasks that are currently arrogated to elite experts, e.g., medical care to doctors, document production to lawyers, software coding to computer engineers, and undergraduate education to professors. My thesis is not a forecast but an argument about what is possible: AI, if used well, can assist with restoring the middle-skill, middle-class heart of the US labor market that has been hollowed out by automation and globalization.

AI can be useful in many ways, as Autor notes. And it's not just "medical care to doctors, document production to lawyers, software coding to computer engineers, and undergraduate education to professors." AIs can be used in virtually any job, offloading the routine parts and allowing us humans to work with the open-ended, context-dependent tasks that AIs don't handle as well (see Chapter 1). Moreover, as we shall discuss in the coming pages, AIs can help each of us to learn skills and knowledge "just in time," as we need them.

The key here is the clause in the last sentence of the above quote: "AI, if used well,…" Ah, but that's the rub. What does it mean to use AIs well? And how can people learn to do this? The following begins to answer these questions.

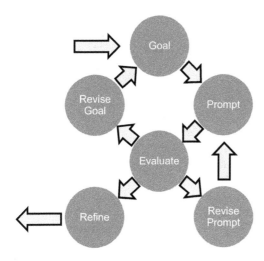

Figure 2.1 The Cognitive Amplifier Loop.

The Cognitive Amplifier Loop

We can get an enormous amount out of AIs as cognitive amplifiers if we engage in a specific set of steps. I describe this process as the *Cognitive Amplifier Loop* (CAL), illustrated in Figure 2.1.

Step 1: Define the Goal

At the outset, we need to be clear on our immediate *goal*, on the reason we are entering something into an AI. We want an AI to either answer a specific question or perform a certain task.

As we begin the CAL, we should ask ourselves the following questions:

- What is the actual question we want to answer or the problem we want to solve? Is it clear and concrete? Should we break it into a set of more precise questions or requests? If we aren't really sure what the question or problem is, perhaps because we are first learning about a subject and aren't very well oriented, we need to take a step back and make *this* the problem we are trying to solve. In this case, we need to specify the goal as trying to distill the problem or question we are struggling to define.
- Can we defend our assumptions about the nature of the problem or issue at hand? Do we seek a black-and-white choice, when in fact there is a messy middle? Are any of our assumptions shaky?

- Is our goal ethical? We need to pause and consider both immediate and long-term ethical consequences of potential solutions, and consider both the desired outcomes and any possible unintended side effects.

For example, let's say that we've read about early "Classical" AI programs, such as IBM's Watson.[3] In 2011 this program won the TV game show *Jeopardy!*, which tests general knowledge. These sorts of computer programs were useful only in very well-defined, limited situations. We might want to know how the new, Generative AIs are different. Thus, our goal might be to have the AI "Summarize briefly the similarities and differences between Generative AI and Classical AI." This is a request for the AI to produce a specific sort of summary, which is concrete. We assume that there are some common features and some different ones; these assumptions are defensible. There are no obvious ethical issues here. This is a very simple goal, which will help us grasp key features of the CAL. Many of the goals we consider later are more complex and nuanced.

Step 2: Create a Prompt

The goal leads us to formulate a specific *prompt*, which is either a question the AI should answer or instructions for the AI to follow to produce a particular product. The guidelines outlined in Table 2.1 can help to create effective prompts.

Several aspects of Table 2.1 are worth underscoring. For one, we can create even very long and complex prompts if we rely on the "Chain of Thought" strategy, which requires us to instruct the AI to go through a logical, step-by-step process.[4] We will see numerous examples of this strategy in later pages. In addition, even though the AI was trained on a vast amount of material, we cannot be certain that it will highlight just those aspects that are relevant in a specific situation. Thus, when in doubt, we should err on the side of being too explicit. It often helps to load in relevant documents and tell the AI to draw on them to answer the question or carry out the request. Furthermore, to make our instructions very concrete and unambiguous, we not only should give the AI examples, but also explain what is important to pay attention to in the examples.[5] Lastly, if there is any room for confusion, we should tell the AI what *not* to do in addition to telling it what to do.

For example, for the goal stated above, "Summarize briefly the similarities and differences between Generative AI and Classical AI," our prompt could be:

Table 2.1 Guidelines for Creating Effective Prompts.

Guideline	Summary	Good Example	Bad Example
Adaptation	Adapt existing prompts by substituting terms, which can save time by minimizing iterative changes.	Using a previous prompt for an event summary and updating it with new event details.	Reusing a scientific research summary prompt for a sports event without modifications.
Unpacking	Break complex goals into distinct parts, number them or use headings, and use a "Chain of Thought" to connect the steps.	"1. Analyze the data. 2. Summarize the results. 3. Suggest inferences from the results."	"Analyze the data and generate insights."
Ambiguity	Avoid vague, abstract, or ambiguous terms. Be literal, expecting the AI to interpret words at their surface meaning.	"Calculate the precise number of new hires needed based on the current turnover rate, expected growth, and budget."	"Figure out how many more people we need."
Assumptions	Make assumptions explicit in the prompt; do not expect the AI to infer them.	"Assuming a 10% annual inflation rate, calculate the future cost."	"Calculate the future cost."
Background	Do not assume that the AI has relevant background information; be explicit and load in necessary documents.	"Using the provided economic data from 2021 in NYC, analyze the trend in consumer spending."	"Analyze the trend in consumer spending in NYC in 2021."
Examples	Provide one or more examples of what is being sought.	If the goal is to have the AI provide scenarios of a healthcare situation that could lead to a conflict, provide one or more examples of such scenarios.	Only note the type of scenarios, without any concrete illustration.
Focus	Make instructions concrete and unambiguous by explaining important aspects of examples and clarifying potential confusions by stating what to do and what not to do.	"List the steps in the chemical reaction, focusing on the temperature and pressure changes. Do not include equations."	"Explain the chemical reaction."

Please provide a comprehensive comparison and contrast between neural-network-driven Generative AI and rules-and-representations-based Classical AI. This comparison-contrast should address both how the AIs are constructed and also how they are used. Regarding uses, note their roles in fields such as education, research, business, and government, and also note potential ethical challenges associated with both or either sort of AI.

Does this prompt respect the list of guidelines we considered? The guidelines in Table 2.1 are a bit much to memorize easily. This sort of situation begs for the CAL, to have an AI relieve us of the cognitive burden of remembering all of this. Thus, I submitted the above draft to GPT-4 with this prompt:

In Part 1 below is a list of considerations for drafting prompts. Could you please use this list to evaluate the prompt in Part 2, and suggest ways I can improve the prompt.
   Part 1. Here are the considerations: [PASTE IN TABLE 2.1]
   Part 2. Here is the draft prompt to evaluate using those considerations: [PASTE IN THE ABOVE PROMPT]

And here is how the AI responded:

AI:

Based on the guidelines listed in Table 2.1 and the draft prompt provided, here's an evaluation with suggestions for improvement:

Evaluation of the Draft Prompt:

Draft Prompt: Please provide a comprehensive comparison and contrast between neural-network-driven Generative AI and rules-and-representations-based Classical AI. This comparison-contrast should address both how the AIs are constructed and also how they are used. Regarding uses, note their roles in fields such as education, research, business, and government, and also note potential ethical challenges associated with both or either sort of AI.

Adaptation

- Current Use: This prompt appears to be original and not adapted from an existing one.
- Suggestion: If a similar comparison has been made previously, consider adapting that prompt to save time and improve clarity.

Unpacking

- Current Use: The prompt broadly covers multiple complex topics but does not break them down into manageable, focused parts.
- Suggestion: Number the sections of the prompt to organize the comparison systematically:

  1. Compare the construction of Generative AI and Classical AI.
  2. Discuss their uses in specific fields.
  3. Outline potential ethical challenges.

Ambiguity

- Current Use: The prompt is relatively clear but uses broad terms like "comprehensive comparison and contrast" which can be interpreted variably.
- Suggestion: Specify what aspects of AI construction and usage should be compared (e.g., underlying algorithms for construction, effectiveness in various fields for usage).

Assumptions

- Current Use: The prompt assumes knowledge about the distinct nature of Generative and Classical AI without specifying what each entails.
- Suggestion: Clarify any technical assumptions or define Generative and Classical AI at the beginning of the prompt.

Background

- Current Use: Assumes that the reader knows how these AIs are applied across different sectors.
- Suggestion: Provide or reference specific examples of how these technologies have been implemented in the mentioned fields, if possible.

Examples

- Current Use: No examples are given, making it harder for the reader to understand the expected depth and scope of comparison.
- Suggestion: Include examples of studies or cases where these AI types have been applied, especially if discussing their roles or ethical implications.

Revised Prompt Example:

"Please provide a detailed analysis of the differences and similarities between neural-network-driven Generative AI and rules-and-representations-based Classical AI:

1. **Construction:** Describe the foundational technologies and algorithms used for each type of AI.
2. **Applications:** Discuss how each AI type is used in education, research, business, and government, providing specific examples or case studies.
3. **Ethical Challenges:** Identify and compare potential ethical issues each AI type may pose in these fields, citing recent controversies or dilemmas."

This revised prompt is more structured, reduces ambiguity, and sets clear expectations for what information needs to be provided.

When I submitted the revised prompt, the results were more detailed and complete than those from the initial prompt. We need to be clear on our actual goal, however, to decide whether we really want a lot of detail. In this case, it was a bit of overkill for me.

But what about the situation where we are not well oriented, and are thrashing around trying to specify the question or instruction? In this case, we take a step back and now ask the AI to produce candidates for the question or instructions. We give it as much information as we can, and then use the results as inspiration—assuming that we can recognize what we seek, even if we cannot state it clearly at the outset. For example:

> Please help me figure out what question I should ask about the nature of computation. I realize that neural-network-driven Generative AI is really different from the kind of computation that underlies word processing. For instance, Generative AI sometimes hallucinates but also can be very creative, coming up with things I didn't program or expect. I also know that traditional computation is less flexible than Generative AI, but it is easier to understand what it's doing. If I want to understand the differences between the different types of computation, what questions should I ask? Please give me five alternatives.

The AI did in fact accommodate. If you are interested in seeing the results, or you want to try any of the other prompts in this book, you can find them at the following website in a form that is easy to copy-and-paste and submit to an AI: www.routledge.com/9781032686660. Unless otherwise noted, all of the interactions with AIs in this book were with OpenAI's GPT-4, which as of this writing is one of the best available LLMs, along with Anthropic's Claude 3.5 Sonnet and Google's Gemini 1.5 Pro.

Step 3: Evaluate the Results

Once the AI has responded to the prompt, we *evaluate* what it produced. During this phase of the CAL, we need to decide whether the results accomplish our goal. The evaluation process requires us to use critical thinking, as summarized in the following chapter. Following our evaluation of the results, one of three things happens. We decide that:

1. The product is good enough to work with, and thus we *retain* it and tune it up "by hand," such as by editing a document or touching up a graphic.
2. Although the product is close to what we want, it's not quite right, and so we go back to Step 2 and *refine* the prompt, and then submit the new version. As soon as the AI has responded to this edited

prompt, we evaluate the new product and consider again whether it's good enough to retain or whether we need to continue to refine the prompt .
3. The product is so far from what we want that we *revise* the goal with a new one and return to Step 1.

With few exceptions, the prompts presented in this book were not the first attempt. In virtually every case, I had to loop through the CAL and revise the prompts to get the sought results. This is often reflected by what I eventually told the AI *not* to do in the prompt, such as not to act on a statement instead of evaluating it.

An example of this process occurred after I saw the initial results from the second, revised prompt described above. I realized that I was in fact especially interested in the ethical implications of Classical vs. Generative AI. Thus, I needed to loop back to the goal (see Figure 2.1) and revise it, which led me to edit the prompt. In this case, I added a single line at the end of the prompt the AI produced above:

> • I am particularly interested in the ethical implications of the two approaches; could you please flesh those out in detail.

When I submitted the revised prompt to GPT-4, it produced the following output:

AI:

Let's delve into a comprehensive comparison and contrast between neural-network-driven Generative AI and rules-and-representations-based Classical AI, with a particular emphasis on their ethical implications.

1. Construction

Neural-Network-Driven Generative AI:

• Architecture: Consists of layers of neurons in a structure inspired by the human brain. This includes input, hidden, and output layers which process data through interconnected nodes.

- Training Methods: Uses large volumes of data and iterative training processes like supervised learning, where the model adjusts based on error rates between actual and predicted outputs.
- Data Requirements: Requires extensive, diverse datasets to effectively learn and generalize. The quality and quantity of data significantly affect the model's performance and bias.

Rules-and-Representations-Based Classical AI:

- Logical Frameworks: Built on logical rules and a symbolic approach, where knowledge is explicitly defined and structured into the system.
- Rule-Based Systems: Operates using if-then rules, making it easier to trace how decisions are made and to ensure compliance with specific regulations.
- Data Handling: Less dependent on data volume but requires meticulously defined rules and accurate data representations to function as intended.

2. Uses

Education:

- Generative AI: Powers adaptive learning platforms that customize content for individual learners and simulate realistic interactions, enhancing engagement and learning outcomes.
- Classical AI: Used in systems that provide consistent instructional content based on fixed rules, beneficial for standardized testing preparation and ensuring uniformity in educational delivery.

Research:

- Generative AI: Facilitates the discovery of new knowledge by identifying patterns and generating insights from large datasets, which can revolutionize fields like genomics and climate science.
- Classical AI: Supports structured research processes, such as data classification and rule-based analysis, providing reliable reproducibility.

Business:

- Generative AI: Enhances customer engagement through personalized marketing and dynamic customer service interfaces.
- Classical AI: Automates routine tasks like inventory management and compliance monitoring, reducing errors and increasing efficiency.

Government:

- Generative AI: Can be used in resource allocation and predictive governance, helping to optimize public services and forecast needs.
- Classical AI: Implements systems in areas such as tax processing and eligibility checks for welfare benefits, ensuring adherence to statutory requirements.

3. Ethical Challenges

Generative AI:

- Bias in Training Data: Reflects existing biases in the input data, potentially leading to unfair or discriminatory outcomes. For example, facial recognition technologies have faced criticism for higher error rates in minority ethnic groups.
- Transparency and Accountability: The complex and opaque nature of neural networks makes it difficult to understand and explain decision-making processes, challenging the principles of transparency and accountability.

Classical AI:

- Rigidity and Lack of Adaptability: The inability to adapt to novel situations not covered by pre-established rules can lead to inefficiencies or incorrect decisions, particularly in dynamic environments.
- Decision-Making Transparency: While decisions in rule-based systems are easier to trace and justify, they can still be opaque when the systems become large and complex, potentially obscuring the logical basis for decisions.

Further Ethical Considerations:

- Long-term Implications: Generative AI's ability to learn and adapt could lead to autonomy in decision-making, raising concerns about the control and misuse of AI technologies. In contrast, Classical AI's rigidity might limit its adaptability to future technological and societal changes, potentially making it obsolete or inappropriate for evolving contexts.
- Regulatory Compliance: Generative AI may require new frameworks for governance to manage its evolving capabilities and impacts, whereas Classical AI fits more comfortably within existing regulatory structures due to its predictable nature.

Focus

The focus remains on highlighting the distinct construction, uses, and particularly the ethical challenges associated with Generative and Classical AI, providing a detailed understanding of their implications in various sectors and their potential long-term effects on society.

I then evaluated this new result from the AI. I decided that I in fact wanted more detail about the ethical implications of the two approaches, and thus looped back to refine the prompt again (see Figure 2.1) while maintaining the same general goal. I entered the below follow-up prompt right after receiving the above result. Because this prompt was in the same "thread" as the original material, I didn't need to repeat the original prompt or provide the results—the AI "remembered" them. Here is my follow-up prompt and the AI's response:

Me:

Thanks. But I would like a more detailed, comprehensive treatment of ethical issues. Could you please expand on that.

AI:

Certainly! Let's dive deeper into the ethical issues associated with both neural-network-driven Generative AI and rules-and-representations-based Classical AI, providing a more detailed exploration of the challenges and considerations.

Ethical Challenges of Neural-Network-Driven Generative AI

Bias and Fairness:

- Generative AI systems, such as those used in hiring tools or predictive policing, often replicate and amplify biases found in their training data. The source of bias can be diverse, ranging from biased historical data, skewed sample representations, to the designers' unconscious preferences. For instance, AI systems in recruitment have been found to favor resumes with names typically associated with men over women, due to historical hiring patterns present in the training data.
- Addressing this requires not only rigorous data sanitization but also continuous monitoring and updating of AI models to ensure fairness and the inclusion of ethical oversight during the development process.

Transparency and Explainability:

- The "black box" nature of deep learning models makes it difficult for users and regulators to understand how decisions are made. This lack of transparency is a major concern in sectors like healthcare or criminal justice, where understanding the basis of AI decisions is critical.
- Efforts to improve explainability include the development of techniques like LIME (Local Interpretable Model-agnostic Explanations) and SHAP (SHapley Additive exPlanations), which attempt to provide insights into the decision-making processes of complex models.

Accountability and Responsibility:

- As AI systems become more autonomous, determining accountability for decisions made by AI becomes challenging. For example, in the case of an autonomous vehicle accident, the question arises whether the manufacturer, the software developer, or the vehicle owner should be held responsible.

- Legal frameworks and guidelines are still evolving to define and enforce accountability, requiring multi-stakeholder collaboration to ensure that laws keep pace with technological advancements.

Privacy:

- Generative AI often requires large amounts of personal data to train effectively. This raises significant privacy concerns, especially when it involves sensitive information. Ensuring that data is used responsibly and that individuals' privacy is protected without stifling innovation is a key challenge.
- Privacy-preserving technologies like differential privacy, federated learning, and secure multiparty computation are being explored to enable the use of data in AI while safeguarding individual privacy.

Ethical Challenges of Rules-and-Representations-Based Classical AI

Rigidity and Contextual Inappropriateness:

- Classical AI systems function based on strict rules, which can lead to outcomes that are inappropriate when context shifts or when rules do not cover unforeseen scenarios. For example, a rule-based diagnostic system might fail in the face of a novel disease that doesn't fit its programmed criteria.
- Ensuring that Classical AI systems are regularly updated and reviewed by human experts can mitigate some of these risks, but it also emphasizes the need for hybrid systems that can incorporate flexibility.

Decision-Making Transparency:

- Although Classical AI is generally more transparent than Generative AI, the logic can still become very complex as the number of rules increases. This can make it difficult to trace how specific decisions are made, potentially leading to errors or misjudgments being hard to detect and correct.

- Clear documentation and regular audits of rule-based systems are essential to maintain their transparency and ensure they operate as intended.

Scalability and Adaptation:

- Classical AI systems are less adaptable to changes than Generative AI, which can be a limitation in environments that require quick adaptation to new data or scenarios. This can lead to outdated or inefficient responses when conditions change.
- Continuous human oversight and the integration of adaptive learning components within Classical AI frameworks can help address these limitations.

Comparative Ethical Considerations

Impact on Society:

- Both types of AI technologies influence societal norms and expectations. Generative AI, with its capability to create realistic simulations and predictions, could potentially blur the lines between reality and simulation, affecting areas such as media, entertainment, and even personal relationships.
- Classical AI, being more predictable and constrained, might reinforce existing societal structures and norms, potentially slowing innovation but providing stability and reliability.

Regulatory Challenges:

- The dynamic capabilities of Generative AI necessitate a more flexible and anticipatory regulatory approach to manage risks without stifling innovation. Conversely, the more static nature of Classical AI allows for more straightforward regulation but may require updates to stay relevant as societal standards evolve.

This expanded discussion aims to provide a deeper understanding of the ethical landscapes of Generative and Classical AI, highlighting the need for thoughtful consideration of these technologies' broader impacts on individuals and society.

This response fell into the first category: good enough to retain and edit by hand.

Step 4: Retain and Edit the Results

To edit the results, I relied on the sorts of critical thinking and creative thinking we discuss in the following two chapters. In those chapters, we review numerous considerations that affect how we edit the results from an AI.

Using the CAL

We can use the CAL for much more sophisticated applications of AI, such as designing "Custom GPTs"—specialized AIs that we design for specific uses—or agents.[6] Indeed, we can instruct the AI to write computer code to perform many different types of tasks, ranging from designing websites to analyzing data. It is tempting to work through such uses of the CAL here, but I did not do so for two reasons: First, this is a moving target. At this writing, the range and reach of AIs is increasing almost on a weekly basis, and it is impossible to list all of the ways we can or could use it. Second, as AIs improve, they will do an increasing number of these tasks without needing explicit instructions. We won't need to call on OpenAI's "Advanced Data Analytics" (previously called "Code Interpreter") to analyze data; we will just need to ask, and the AI will carry out the details. We won't need to describe the code we want the AI to write; we will only need to carefully specify the goal, and the AI will figure out the code for us. And so on.

In all cases, present and future, the CAL will serve us well. The process outlined in the CAL is much more efficient than starting with an ill-defined goal, not designing an appropriate prompt to address the goal, not deliberately evaluating the AI's product in the context of the goal, and so on. Using this CAL process can help us achieve our goals more effectively and efficiently than a haphazard, purely trial-and-error approach.

However, we need to be aware of a potential problem with using the CAL: We can get sucked in and end up spending an inordinate amount of time cycling through it with continued refinements (or what feels like refinements). This occurs because of "partial reinforcement effects."[7,8,9]

That is, people tend to repeat a behavior that leads to a desired outcome. This outcome is called "positive reinforcement" for that behavior. People are particularly prone to persist in the behavior when the reinforcement does not occur every time. If the reinforcement occurs only some percentage of the time, we tend to keep plugging away, seeking that reinforcement. Thus, if every few cycles of the CAL produces even a partially useful result, getting us closer to our goal, we typically will persist in trying—even (speaking from personal experience) deep into the night. We need to be aware of this, and be prepared to impose a pause on our efforts after a reasonable amount of time. Indeed, pausing can result in "incubation," where our unconscious works on the problem and can produce solutions that we had not previously considered.[10,11]

Do We Need Humans in the Loop?

Given how well AIs can analyze, evaluate, and create text and images, it might be tempting to question why humans need to evaluate results from an AI. Why not just have the AI itself engage in the necessary critical thinking? Or perhaps we could have two instances of the system running, with one evaluating the products of the other. Consider four reasons why humans must be in the loop.

First, even very advanced AIs are unlikely to be perfect, in part because of their neural net design (as discussed in the previous chapter) and also because they will never be trained on perfect data sets. Their flaws not only affect their initial response to a prompt, but also affect everything else they do. If we asked one AI to evaluate the results from another, we would then need yet another AI to evaluate the results from the AI that was doing the evaluation, and then yet another to evaluate the results from that last one, and so on, in infinite regress. We humans are also not perfect, of course, but our errors typically won't go against our interests—at least not for very long. In contrast, the AI's errors may be random or biased in unknowable ways, which means that at least some of the time they may be against our interests (cf. [12]). We need humans in the loop to act as gatekeepers, which is analogous to how academics serve as gatekeepers for journals: they rely on what they already know to evaluate the quality of a product. We consider how to do this in the following chapter.

Second, an AI is inert until prompted. Each individual human user sets the goal, which then guides the prompts, evaluations, and eventual final editing of the results. True, we can ask even today's AIs to formulate goals, and—as illustrated above—we can ask them to develop prompts based on those goals. But that is beside the point. If an AI were to set goals from scratch, they wouldn't be *our* goals. If a human is working with an AI as a cognitive amplifier, by definition a human is going to need to be involved in setting its goals, designing the prompts, evaluating the results, and tuning up the end product.

Third, because an AI can never know absolutely everything about each and every user, it can never be certain what the user is actually seeking at any one moment in time. Hence, it cannot develop optimal goals for each of us. As noted in the previous chapter, part of the problem is that an AI can never be completely aware of the context surrounding any individual human being. AIs cannot be trusted to make decisions autonomously for us.

And fourth, humans must develop their own goals in part because they often may not consciously know in advance what their precise goals are. The US Supreme Court justice Potter Stewart in 1964 famously refused to define pornography, and instead simply asserted that he "knows it when he sees it." People ridiculed him at the time, which made sense in the context of the law: Standards for obscenity should be objective, given that people are expected to comply with them. But this sort of response is in fact reasonable in the context of developing our goals when we work with AIs. Our ability to recognize usually outstrips our ability to recall or produce. For example, most people who are familiar with a second language can understand more than they can say, which highlights the distinction between recognizing what comes in vs. being able to produce something. Users of the CAL may produce the best approximations they can to their goal, and iterate until they recognize the sought product when they see it. Not knowing everything about any given human, an AI cannot do this.

In short, the human user cannot be replaced by an AI. An AI is a cognitive amplifier, and it needs human input to spark it to life. Humans thus need to learn to deploy AIs effectively, efficiently, and ethically.

Flourishing with the CAL

In the previous chapter, we noted that learning to interact effectively with AIs is a critical skill to flourish in the Age of AI. Acquiring this skill can produce a sense of autonomy and control. It can also lead us to be more satisfied at work, especially if we learn how to be more productive using AIs and learn how to work with them to produce high-quality products. Developing such skills can also help us have job stability because we can be more proactive, and thus become a valued employee.

To master working with AIs, we need to know how and when to initiate interactions with them, how to interpret the AI's claims, products, and suggestions, and how to steer AIs in directions we desire. The CAL is a response to these needs, and provides a structure for interacting with AIs. If we master the CAL, we will nail down one of the requirements for flourishing in the emerging world.

To use the CAL effectively, we need to engage in critical thinking. In fact, the CAL can lead us to be better critical thinkers, both by helping us directly and by training us to be better at such thinking. Critical thinking is absolutely essential in the Age of AI, not only to use the CAL, but also to sort through the chaff being produced when others employ the CAL to create misinformation and disinformation. We will never be able to stop all such material, and hence it's important to learn how to employ AIs to help us spot them and respond appropriately. We consider critical thinking in the following chapter.

Digging Deeper

For readers who want to dig deeper into the subject matter, below are examples of videos that provide accessible, more in-depth treatment of aspects of the material reviewed in the chapter. In addition, I've also included search terms that can help the reader locate new videos on the topic.

What is Cognitive AI? Cognitive Computing vs Artificial Intelligence | AI Tutorial | Edureka: https://www.youtube.com/watch?v=Zsl7ttA9Kcg

Prompt Engineering Course. How To Effectively Use ChatGPT & Other AI Language Models: https://www.youtube.com/playlist?list=PLYio3GBcDK sPP2_zuxEp8eCulgFjl5a3g

AI in a Minute: Prompt Engineering: https://www.youtube.com/watch?v=vGdyePbGNaE

- "Augmented intelligence and human–AI collaboration"
- "Using AI to enhance human cognitive strengths"
- "AI compensating for human cognitive weaknesses"
- "Leveraging AI for decision-making support"
- "AI as a tool for extending human capabilities"
- "Designing clear and concise AI prompts"
- "Evaluating AI-generated outputs"
- "Iterative refinement of AI prompts"
- "Adapting AI prompts for similar tasks"

Don't fear intelligent machines. Work with them | Garry Kasparov: https://www.youtube.com/watch?v=NP8xt8o4_5Q

The limits of AI and ChatGPT: The common sense problem: https://www.youtube.com/watch?v=w9MSsnoTSJs

- "Human-in-the-loop AI"
- "Importance of human oversight in AI decision making"
- "AI limitations in understanding human goals and context"
- "Human–AI collaboration in goal setting"
- "AI bias and the need for human evaluation"
- "Iterative goal refinement with AI assistance"
- "Ethical considerations in human–AI interaction"
- "Developing effective human–AI partnerships"
- "Human cognitive limitations and AI amplification"

References

1   Mitchell, A. (2022, December 26). Professor catches student cheating with ChatGPT: "I feel abject terror." *New York Post*. https://nypost.com/2022/12/26/students-using-chatgpt-to-cheat-professor-warns/

2   Autor, D. (2024). *Applying AI to rebuild middle class jobs*. National Bureau of Economic Research Working paper 32140.

3   Walsh, B. (2021, February 13). Looking back at Watson's 2011 "Jeopardy win." *Axios Technology*. www.axios.com/2021/02/13/ibm-watson-jeopardy-win-language-processing

4   Kojima, T., Gu, S. S., Reid, M., Matsuo, Y, & Iwasawa, Y. (2023, January 29). *Large language models are zero-shot reasoners*. Cornell University. https://arxiv.org/abs/2205.11916

5   Mollick, E. (2023, August 13). *Automating creativity*. www.oneusefulthing.org/p/automating-creativity

6   Mollick, E. (2024, May 14). *What OpenAI did: A new model opens up new possibilities*. www.oneusefulthing.org/p/what-openai-did

7   Hochman, G., & Erev, I. (2013). The partial-reinforcement extinction effect and the contingent-sampling hypothesis. *Psychonomic Bulletin and Review*, *20*, 1336–1342. doi:10.3758/s13423-013-0432-1.

8   Kosslyn, S. M., and Rosenberg, R. S. (2019). *Introducing psychology: Brain, person, group* (5th ed.). Flatworld.

9   Mellgren, R. L. (2012). Partial reinforcement effect. In N. M. Seel (Ed.), *Encyclopedia of the sciences of learning*. Springer. https://doi.org/10.1007/978-1-4419-1428-6_276

10 Baird, B., Smallwood, J., Mrazek, M., Kam, J. W., Franklin, M. S., & Schooler, J. W. (2012). Inspired by distraction: Mind wandering facilitates creative incubation. *Psychological Science*, *23*, 1117–1122. https://doi.org/10.1177/0956797612446024.

11   Talandron-Felipe, M. M. P., & Rodrigo, M. M. T. (2021). The incubation effect among students playing an educational game for physics. *Research and Practice in Technology Enhanced Learning, 16*. https://doi.org/10.1186/s41 039-021-00171-x

12   Koos, R., Lee, M., Raheja, V., Park, J. I., Kim, Z. M., & Kang, D. (2023). *Benchmarking cognitive biases in large language models as evaluators*. Cornell University. https://arxiv.org/abs/2309.17012

Chapter 3

Boosting Critical Thinking

Critical thinking, at its core, is about analyzing and evaluating, which ultimately leads to a judgment or decision. We need to engage in critical thinking at every step of the Cognitive Amplifier Loop (CAL), and critical thinking can help us in many other, human-oriented, contexts. Without question, we are going to need to become sophisticated critical thinkers in order to flourish in the Age of AI.

A key insight about critical thinking is that it is not "one thing." Rather, it has many facets, as is evident in the following excerpt of a definition of the term.[1]

> Critical thinking is the intellectually disciplined process of actively and skillfully conceptualizing, applying, analyzing, synthesizing, and/ or evaluating information gathered from, or generated by, observation, experience, reflection, reasoning, or communication, as a guide to belief and action. In its exemplary form, it is based on universal intellectual values that transcend subject matter divisions: clarity, accuracy, precision, consistency, relevance, sound evidence, good reasons, depth, breadth, and fairness.
>
> It entails the examination of those structures or elements of thought implicit in all reasoning: purpose, problem, or question-at-issue; assumptions; concepts; empirical grounding; reasoning leading to conclusions; implications and consequences; objections from alternative viewpoints; and frame of reference...

In this chapter, we break down critical thinking into distinct types that are relevant to the CAL and to living in an AI-infused world.

DOI: 10.4324/9781032686653-4

Using the CAL to Enhance Critical Thinking

Table 3.1 summarizes relevant types of critical thinking. This table presents a lot of information, which is difficult for us mortals to absorb. In such situations, we can engage an AI as a cognitive amplifier, helping us manage our cognitive limitations. We can use the CAL to offload a lot of the work required to remember and apply the various types of critical thinking.

We can ask an AI to draw on Table 3.1 to help us engage in critical thinking. We could simply rely on the training data that went into the AI, but drawing on the table provides more transparency into what the AI is actually doing. For example, I gave the following prompt to GPT-4o:

Please begin by asking me to give you a description of a situation that requires critical thinking. I want you to help me evaluate this situation. Wait for me to describe the situation before continuing. Only ask me additional questions after I have finished writing about the situation. After I give you the situation, please do the following:

1. Ask me questions one at a time that you need to have answered in order to use the techniques in the attached Table to make a judgment. Ask me one question at a time, and wait until I respond before asking me another question. Do not continue until I have answered each of your questions, one at a time.

2. After asking me whatever questions you need answered, please give me advice. What distinctions should I make? What should I be cautious about? What questions should I ask? After you have offered your advice, please ask whether I have any additional questions or requests, and respond appropriately.

3. At the end of the session, please provide a summary that indicates which techniques from the attached Table you used and how you used them. Attached is the Table that should guide your critical thinking. [ATTACH TABLE 3.1]

What follows are the results from my submitting the above prompt along with Table 3.1.

Table 3.1 Types of Critical Thinking Relevant for the CAL.

Category	Class	Type	Explanation	Example
Define the Problem	Identifying the Actual Problem	Decomposing Complex Problems	Breaking down complex issues into smaller, manageable parts.	Analyzing a complex project by separating it into milestones, tasks, and resources required.
		Vagueness and Ambiguity	Clarifying unclear or double-meaning statements for precise understanding.	Refining a goal from "improve customer satisfaction" to "increase customer survey scores by 10%."
	Questioning Assumptions	Misleading Dichotomies	Identifying false binaries in arguments, suggesting oversimplification.	Debating the "individual effort vs. systemic factors" in educational achievement.
		Inappropriate Assumptions	Challenging assumptions that may not be valid or evidence-based.	Examining the assumption that increased screen time directly causes decreased physical activity in children.
	Evaluating Ethical Implications	Short-Term vs. Long-Term	Weighing immediate benefits against long-term impacts.	Considering the ethical implications of using non-renewable energy for short-term gains versus investing in renewable energy for long-term sustainability.
		Intended vs. Unintended Consequences	Considering both the desired outcomes and potential unintended side effects.	Evaluating the introduction of an invasive species for pest control and its unintended effects on the ecosystem.

(Continued)

Table 3.1 (Continued)

Category	Class	Type	Explanation	Example
Evaluate Data and Claims	Considering the Source	Source Intent	Understanding the purpose behind the information provided.	Distinguishing between an informative scientific report and a company's promotional material.
		Source Expertise	Considering the knowledge and expertise of the information provider.	Distinguishing health guidelines from a public health agency versus general advice from a non-medical online forum.
	Establishing Differences and Trends	Statistical Significance	Establishing that two measures are actually different.	Demonstrating that active learning really does produce better learning than traditional methods.
		Characterizing Trends	Fitting straight lines or curves to data.	Showing that sales really are increasing over time.
	Evaluating Evidence	Lumping or Splitting	Deciding whether to group information together or analyze it separately.	Separating symptoms into categories for a more accurate medical diagnosis.
		Relevance	Judging the pertinence of information.	Focusing on recent studies about climate change and ignoring outdated, disproven theories.
Analyze Inferences	Evaluating Arguments	Form: Logical Consistency	Ensuring that arguments are free from contradictions and logically coherent.	Pointing out contradictions in an article arguing for and against tax increases in the same breath.
		Content: True Premises	Verifying that the premises of an argument are accurate and supported by evidence.	Fact-checking a news article's claims against credible data sources.

Table 3.1 (Continued)

Category	Class	Type	Explanation	Example
	Evaluating Causality	Correlation vs. Causation	Distinguishing between relationships where one event causes another and where they merely occur together.	Analyzing whether increased ice cream sales lead to higher rates of sunburn.
		Missing Control Condition	Identifying when comparisons lack a control group for accurate assessment.	Critiquing a study on sleep patterns without a control group for comparison.
		Multiple Causes	Considering that an outcome may have several contributing factors.	Investigating economic downturns by examining political, social, and technological factors.
Make Decisions	Defining Options	Trade-Offs	Weighing the pros and cons of each option.	Listing benefits and drawbacks of different investment strategies before choosing one.
		Effects of Emotion	Recognizing how emotions influence decision making and striving for objectivity.	Acknowledging the impact of fear in emergency decision making and seeking calm analysis.
	Assessing Feasibility	Technical	Assessing the technical viability of a solution.	Evaluating the feasibility of implementing a new software platform within current IT infrastructure.
		Operational	Determining whether a solution can be implemented within the operational constraints.	Assessing the compatibility of a new workflow with existing team capabilities and resources.
		Financial	Evaluating whether the organization can afford the solution.	Assessing the total cost of implementing a new computer system and training employees how to use it.

AI:

Please describe a situation that requires critical thinking. I'll help you evaluate it step by step using the techniques from the attached Table 3-1.

Me:

I heard a colleague claim that he could create a free university that would provide high-quality education for all Americans, and graduates would have an advantage in being hired for federal jobs. Should I believe this?

AI:

Thank you for sharing this situation. I will help you evaluate it using the techniques from the attached Table 3-1. To start, I need to ask you a few questions to gather more information.

Question 1:
What specific assumptions does your colleague make to support his claim that he can create a free university providing high-quality education for all Americans?

Me:

He is assuming that he can train people in such useful knowledge and skills that employers would pay the university to hire the graduates. Moreover, he assumes that students would recognize the value of this and sign up in large numbers.

AI:

Thank you for your response. Let's delve deeper into these assumptions.

Question 2:
Has your colleague provided any evidence or examples to support the feasibility of employers paying the university to hire its graduates?

Me:

No, he has not.

AI:

Question 3:
Has your colleague provided any details on how he plans to attract a large number of students to this free university? For instance, has he mentioned any marketing strategies or partnerships that would ensure significant enrollment?

Me:

Yes. He plans to write jingles that go with cute dances, and he will post these on TikTok. He also plans to try to partner with large companies, such as Amazon and Starbucks—but he hasn't talked to them yet to find out whether they are interested.

AI:

Question 4:
Has your colleague outlined any specifics about the curriculum and how it ensures high-quality education? For example, has he mentioned accreditation, faculty qualifications, or any innovative teaching methods?

Me:

His idea is a bit radical. He wants to analyze the job descriptions from employers, and indicate what skills and knowledge are necessary to fill each of the jobs. He then wants to provide evidence that graduates have those specific skills and knowledge. His idea is that employers will care much more about this than accreditation, so he isn't going to pursue it.

AI:

Question 5:
Has your colleague addressed the potential financial and operational feasibility of running such a university, especially considering it will

be free for students? For instance, how does he plan to cover operational costs, faculty salaries, and other expenses?

Me:

Yes. He thinks that employers will be so motivated to hire graduates that it won't be necessary to charge tuition or obtain federal funds. Furthermore, his idea is to have a third-party, non-profit, certify that the graduates do in fact have the claimed skills and knowledge. He thinks that certification will take the place of accreditation in most cases.

AI:

Question 6:
Has your colleague provided any plan or evidence to address the ethical and long-term implications of his university model? For instance, the potential impact on traditional universities, student outcomes, and the job market?

Me:

No, he has not.

AI:

Based on the information provided, here is some advice and considerations:

Distinctions to Make:

1. **Assumptions vs. Evidence**: Your colleague makes several assumptions without concrete evidence, such as employer willingness to pay for hiring graduates and large-scale student enrollment due to social media marketing.
2. **Accreditation vs. Certification**: The distinction between traditional accreditation and third-party certification needs careful evaluation. Accreditation ensures quality standards, while certification might not be universally accepted.

Cautions:

1. **Feasibility**: Relying solely on employer funding is risky without prior commitments or evidence that employers will find this model appealing and sustainable.
2. **Operational Costs**: Running a free university involves significant expenses. Without a clear financial plan, this model may face serious challenges.
3. **Ethical Implications**: The long-term impact on the education system and job market must be considered. Dismissing traditional accreditation might undermine the perceived credibility of the institution.

Questions to Ask:

1. What specific evidence supports the assumption that employers will fund the university based on graduates' skills and knowledge?
2. How will the university ensure the quality and consistency of education without traditional accreditation?
3. What contingencies are in place if partnerships with large companies or mass student enrollment do not materialize?
4. How will the university handle the potential ethical and long-term impacts on traditional universities and the broader educational ecosystem?

Techniques From Table 3-1 Used:

1. **Questioning Assumptions** (Misleading Dichotomies, Inappropriate Assumptions): Evaluating the assumptions about funding and accreditation.
2. **Evaluating Ethical Implications** (Short-Term vs. Long-Term, Intended vs. Unintended Consequences): Considering the broader impact on education and society.
3. **Evaluating Evidence** (Source Expertise, Relevance): Assessing the credibility and pertinence of the proposed model and strategies.

Do you have any additional questions or requests?

Me:

No, thank you.

AI:

You're welcome! Here's a summary of the techniques from Table 3-1 that we used and how they were applied:

Summary of Techniques Used:

1. **Questioning Assumptions**:
   - **Misleading Dichotomies**: We examined the assumption that certification could entirely replace accreditation without considering potential complexities.
   - **Inappropriate Assumptions**: We questioned the assumption that employers would fund the university and that social media marketing would guarantee student enrollment.

2. Evaluating Ethical Implications:
   - **Short-Term vs. Long-Term**: We considered the immediate feasibility versus the long-term sustainability and impact on the education system.
   - **Intended vs. Unintended Consequences**: We discussed potential unintended consequences such as undermining traditional universities and the broader educational ecosystem.

3. **Evaluating Evidence**:
   - **Source Expertise**: We evaluated the credibility of the proposed model and the lack of concrete evidence supporting the assumptions.
   - **Relevance**: We assessed the pertinence of the colleague's strategies and plans in the context of creating a high-quality, free university.

The AI was helpful, prompting me to think critically about relevant issues. However, some of this seemed a bit superficial to me, and it would be worth a second cycle of the CAL to push the AI to go deeper. In addition,

I was a little surprised by the redundancy at the end of its response, where it indicated which techniques from Table 3.1 it employed. But this is actually a feature, not a bug: A byproduct of asking the AI to indicate how it drew on the techniques in Table 3.1 is that we will become familiar with the various forms of critical thinking over time, and at some point we can take off the "training wheels"—we will not always need to rely on the AI. Having the AI hammer home the techniques it used will help us learn the techniques.

Indeed, by asking the AI to specify the types of critical thinking it drew on, we can address the single most difficult problem in the science of learning, namely the problem of transfer: We don't automatically apply what we learn in one context to other contexts.[2,3,4] The most effective way to address this problem is to consider many and varied examples of how to apply a principle and ensure that we deeply grasp what those examples have in common. Just being exposed to the different types of critical thinking is not enough; we need to learn how to apply them in many situations.

Critical Thinking for the CAL

We just considered how to use the CAL to help us humans to think critically. Such critical thinking is crucial at every step of the CAL.

To reiterate, the CAL has the following steps:

1. Setting the Goal
2. Creating a Prompt
3. Evaluating the Results
4. Optional Looping Back either to Step 1 (Revise the Goal) or to Step 2 (Refine the Prompt)
5. Retaining and Editing the Results

Let's consider each step in turn, and see how critical thinking helps us use the CAL—and see how an AI can employ critical thinking to help us use the CAL effectively and efficiently.

Setting the Goal

We noted previously that we may sometimes be a bit hazy about what we are seeking when we engage an AI. In such straits, we can use the CAL to sharpen our goal. Indeed, we may not always realize when our goal is hazy or vague. For instance, we might decide that we "want to exercise more," and don't realize how vague that goal is. An AI can sharpen our goals by deploying the SMART criteria.[5,6] SMART goals are *Specific*, *Measurable*, *Achievable*, *Relevant*, and *Time-bound*:

- *Specific.* The goal should be clear and detailed. As already noted, we need to state exactly what we want to accomplish.
- *Measurable.* We need to formulate concrete criteria to measure progress toward achieving the goal, otherwise we won't know when we've actually achieved the goal.
- *Achievable.* The goal should be realistic, given the resources we have available. We shouldn't set ourselves up for failure.
- *Relevant.* A short-term goal of the sort we enter into the CAL should align with and support our larger, more long-term objectives and priorities.
- *Time-bound.* We need a clear sense as to how long it should take to achieve the goal, and work to achieve that self-imposed deadline.

Here's a prompt we can use to ask an AI to deploy the SMART criteria to evaluate a goal, using the AI as a cognitive amplifier:

> Please ask me to submit a statement of a goal. Wait for me to do so before continuing. Don't do anything more until I've provided a statement of the goal. This statement will not be a request for you to do anything, so please do not respond by answering the statement. Rather, the statement will be something that you need to evaluate; it is not a request. After I've provided the goal statement, evaluate that statement using the SMART criteria. Do not respond to the goal itself, but treat it as a statement that you will evaluate. Specifically, consider how the goal statement aligns with being Specific, Measurable, Achievable, Relevant, and Time-bound. Give me feedback and constructive suggestions. Following this, ask me if I want to update the statement of the goal. If I do, ask me for the new goal statement and then evaluate it using the SMART criteria. Please continue doing this until I am satisfied with the goal statement.

Here's the cycle that refined my goal according to the SMART criteria:

> **AI:**
>
> Please submit a statement of a goal.
>
> **Me:**
>
> Identify the ways that generative AI can help people when they want to learn specific skills or knowledge.

AI:

Your goal statement is: "Identify the ways that generative AI can help people when they want to learn specific skills or knowledge."

Evaluation using SMART criteria:

1. **Specific:** The goal is somewhat broad as it does not specify which skills or knowledge areas are of interest or what aspects of learning are being targeted (e.g., comprehension, retention, application). It could be more specific by narrowing down the type of skills, knowledge, or learning scenarios.
2. **Measurable:** The goal does not establish clear criteria for measurement. It could be improved by defining what constitutes "identification" of ways generative AI can help. For instance, a measurable goal might include identifying a certain number of methods or documenting specific case studies.
3. **Achievable:** The goal seems achievable as generative AI has well-documented applications in learning and education, and there are likely multiple resources and studies available on this topic.
4. **Relevant:** The goal is relevant, especially in the context of educational technology and AI's increasing role in personalized learning environments.
5. **Time-bound:** The goal lacks a time frame. Adding a deadline by which the ways should be identified would make it more effective. For example, "within the next three months."

Suggestions for improvement:

To make your goal statement more effective according to the SMART criteria, consider specifying the type of skills or knowledge areas, how you will measure the identification process, and setting a clear deadline. For instance, "Identify at least five ways generative AI can assist in learning mathematical skills for high school students by the end of Q3 2024."

Would you like to update the statement of the goal? If so, please provide the new goal statement.

Me:

Yes, how about: "Identify at least 6 ways that generative AI can help anyone learn to use AI effectively."

AI:

Your revised goal statement is: "Identify at least 6 ways that generative AI can help anyone learn to use AI effectively."

Evaluation using SMART criteria:

1. **Specific:** The goal is specific by targeting the identification of ways generative AI can assist in learning to use AI effectively. However, the term "anyone" could be more specific. Are you referring to students, professionals, or a general audience?
2. **Measurable:** The goal is measurable as it specifies the identification of at least 6 ways.
3. **Achievable:** This goal is achievable, considering the increasing resources and examples of how generative AI is used in educational contexts to teach AI concepts and usage.
4. **Relevant:** The goal is highly relevant, especially as AI continues to be a significant area of interest and its understanding becomes increasingly essential across various fields.
5. **Time-bound:** The goal currently lacks a specific deadline. Including a time frame would make this goal complete and more compelling. For example, "by the end of the year" or "within the next six months."

Suggestions for improvement:
To refine your goal further, consider specifying the target audience more precisely and adding a deadline to enhance its time-bound quality. For instance, "Identify at least 6 ways that generative AI can help students learn to use AI effectively by December 2024."

Would you like to update the statement of the goal again? If so, please provide the new goal statement.

Me:

Thank you.

AI:

You're welcome! If you need any more help in the future, feel free to ask. Have a great day!

Enlisting an AI to think critically about the goal was helpful. It applied the SMART criteria in reasonable ways and nudged me to make the goal more specific and concrete. However, I decided that I didn't need to specify the nature of the learners or a timeframe. I want to cast the net very widely, and my previous experience with similar projects allowed me to assess the feasibility of this one quickly and easily.

In sum, when we first conceive of the goal for engaging the CAL, we can save time by engaging in critical thinking—and having an AI help us to do so. This approach helps us avoid false starts, rabbit holes, and much confusion.

Creating a Prompt

The next step is to create a prompt that addresses the goal. Again, an AI can employ critical thinking to sharpen this process. For example, let's pursue the goal "Identify at least 6 ways that generative AI can help anyone learn to use AI effectively." Here's a first draft of a prompt that addresses that goal:

Please conduct a comprehensive evaluation of the use of Generative AI in helping all kinds of people to learn skills and knowledge more effectively. This evaluation should include a balanced discussion of the benefits and drawbacks of using AI in various capacities in both formal and informal learning environments.

We can now ask the AI to draw on Table 3.1 and Table 2.1—which specifies criteria for effective prompts—to refine the draft prompt:

I am going to give you a draft of a prompt I want later to submit to you. Please evaluate this prompt according to the criteria in the attached Tables, and suggest a better way to write it, given my goal. In what follows I first give you my goal and then I give you the prompt itself.

1. My goal is: "Identify at least 6 ways that generative AI can help anyone learn to use AI effectively."
2. Draft prompt: "Please conduct a comprehensive evaluation of the use of Generative AI in helping all kinds of people to learn skills and knowledge more effectively. This evaluation should include a balanced discussion of the benefits and drawbacks of using AI in various capacities in both formal and informal learning environments."
3. Critical thinking criteria: The criteria I would like you to use to improve the prompt are in the attached Tables. [ATTACH TABLES 2.1 and 3.1]

Here's how GPT-4 responded:

Based on the critical thinking criteria and guidelines provided in Tables 2.1 and 3.1, here are some improvements and restructurings for your draft prompt:

Original Draft Prompt: "Please conduct a comprehensive evaluation of the use of Generative AI in helping all kinds of people to learn skills and knowledge more effectively. This evaluation should include a balanced discussion of the benefits and drawbacks of using AI in various capacities in both formal and informal learning environments."

Revised Prompt: "Identify and describe at least six specific ways in which generative AI can assist diverse individuals in learning to use AI tools effectively. For each method:

1. Detail the practical application in both formal education settings (like schools) and informal learning environments (like online tutorials).

2. Discuss the direct benefits of this approach.
3. Consider any potential drawbacks or limitations.
4. Provide concrete examples to illustrate each method where possible."

Justification for the Revision:

1. **Unpacking and Decomposing Complex Problems:** The original prompt asked for a "comprehensive evaluation," which is broad and complex. The revised prompt breaks down the request into specific, manageable parts: identifying ways, detailing applications, discussing benefits and drawbacks, and providing examples.
2. **Avoiding Ambiguity:** The original prompt used vague language such as "helping all kinds of people" and "more effectively," which can be interpreted in many ways. The revised prompt specifies what is meant by "effectively" by asking for practical applications, benefits, and drawbacks.
3. **Making Assumptions Explicit:** The revised prompt clearly states the expectation of discussing both formal and informal settings, making it explicit where the AI's applications should be considered.
4. **Providing Examples:** By asking for concrete examples, the revised prompt aligns with the guideline to provide examples of what is being sought, enhancing clarity and understanding.
5. **Adapting the Language:** Instead of a broad evaluation, the revision focuses on identifying and describing methods, which is more targeted and likely to yield actionable insights.

This revision should help in generating a more structured, clear, and practical response from the AI, making it easier for users to understand and apply the information in real-world contexts.

I wasn't satisfied with this response because it didn't draw enough on the sorts of critical thinking summarized in Table 3.1, and so decided to iterate. In the same thread, I entered the following:

Me:

You've done a nice job of applying the material in Table 2.1, but I'm wondering whether more critical thinking—of the sorts outlined in Table 3.1—would be helpful. Could you please take another look, with more critical thinking?

AI:

Certainly! Incorporating more critical thinking elements from Table 3.1 can further refine the prompt. Let's revisit the prompt with an added emphasis on critical thinking aspects such as questioning assumptions, evaluating ethical implications, and analyzing inferences.

Enhanced Prompt with Critical Thinking Elements: "Identify and critically assess at least six ways generative AI can facilitate learning to use AI tools effectively for diverse users. For each method, please address the following:

1. **Detailed Application:** Describe how the method can be applied in both formal educational settings and informal learning environments. Include assumptions about user access to technology and prior knowledge.
2. **Benefits Analysis:** Discuss the direct benefits of each method, citing evidence or studies where available.
3. **Drawbacks and Ethical Considerations:** Identify potential drawbacks and ethical implications, such as data privacy concerns or the potential for reinforcing biases.
4. **Concrete Examples and Case Studies:** Provide concrete examples or case studies that illustrate each method.
5. **Evaluating Causality and Relevance:** Analyze whether the success attributed to these methods is genuinely due to AI or if other factors are also at play. Evaluate the relevance of the examples to current technology and learning theories.
6. **Trade-offs and Decision Making:** Discuss the trade-offs involved in implementing each method, considering factors like cost, scalability, and inclusivity."

Justification for Enhanced Prompt:

1. **Questioning Assumptions:** The prompt now explicitly asks for considerations about user access and prior knowledge, challenging the assumption that all users start from the same level or have the same resources.
2. **Evaluating Ethical Implications:** It introduces the need to discuss ethical issues, such as privacy and bias, which are critical when deploying AI in learning environments.
3. **Evaluating Causality and Relevance:** This part of the prompt asks to differentiate between causation and correlation, ensuring that the benefits attributed to AI are not confused with other variables.
4. **Trade-offs and Decision Making:** This addition encourages a deeper analysis of the practicalities of implementing the AI methods, looking at both the benefits and the costs or potential downsides.

These enhancements should deepen the level of critical thinking in the responses generated, leading to a more nuanced and comprehensive evaluation of how generative AI can aid in learning.

And the AI did indeed produce a much fuller, more thoughtful and nuanced response after I submitted that revised prompt. Here's a sample of the output, which I truncate in the interests of space:

1. Personalized Learning Experiences

Detailed Application: Generative AI can tailor educational content based on the learner's pace, preference, and performance. In formal settings like schools, AI can integrate with existing digital platforms to personalize lesson plans and assessments. In informal settings, such as online self-learning platforms, AI can recommend resources and activities to suit the user's unique learning path. This assumes users have basic access to technology and minimal prior knowledge of AI.

Benefits Analysis: Personalized learning using AI has been shown to increase engagement and improve learning outcomes, as it addresses individual learning needs and styles. For instance, a study published in the "Journal of AI Research" highlighted improved test scores in courses where AI personalized learning paths.

Drawbacks and Ethical Considerations: A significant concern is data privacy, as personalization requires collecting detailed data on users' learning behaviors. Additionally, if not carefully designed, AI systems could reinforce biases present in the training data, leading to unequal learning opportunities.

Concrete Examples: Carnegie Learning's "MATHia" software uses AI to provide personalized math instruction, adapting to students' responses in real time.

Evaluating Causality and Relevance: While improvements in learning outcomes are often attributed to AI, the quality of the underlying educational content and the educator's role in facilitating learning are also crucial. The relevance is significant as personalized learning aligns with contemporary educational theories that emphasize learner-centered education.

Trade-offs and Decision Making: Implementing AI for personalization requires significant investment in technology and training for educators. While it offers scalability, the cost and need for continuous updates and monitoring to avoid biases are considerable challenges.

2. Simulation and Modeling Tools

Detailed Application: Generative AI can create realistic simulations and models that allow learners to experiment and learn in a risk-free environment. In formal education, such simulations can be part of the curriculum in subjects like science and engineering. Informally, enthusiasts and professionals can use these tools to hone their skills or understand complex systems. This method assumes moderate to high access to technology.

Benefits Analysis: Simulations facilitate experiential learning, which is known to enhance comprehension and retention. Studies in "Educational Psychology" have shown that students using simulations perform better in problem-solving assessments.

Drawbacks and Ethical Considerations: High-quality simulations require significant computational resources, which might not be accessible to all. Moreover, there's a risk of over-reliance on simulations, potentially overlooking practical skills.

Concrete Examples: Flight simulators used in pilot training and Foldit, a puzzle game about protein folding, are examples where AI-driven simulations aid learning significantly.

Evaluating Causality and Relevance: The effectiveness of simulations in enhancing learning is well-supported by cognitive theory, which posits that active engagement in learning activities promotes deeper understanding. However, the technological sophistication also plays a critical role.

Trade-offs and Decision Making: The decision to implement simulation tools involves balancing the cost of development and maintenance with the potential for wide-reaching, impactful learning experiences.

[TRUNCATED]

Clearly, an AI can help us develop prompts. As illustrated, we can engage the CAL iteratively, having the AI use critical thinking to refine the prompt.

Evaluating the Results

The result at the end of the previous section was better than what I got when I submitted the original prompt. It is more complete and "thoughtful." However, this is a subjective judgment. We can do better than that by using an AI as a cognitive amplifier. We can induce an AI to employ many types of critical thinking in the "Evaluation" phase of the CAL. For example, here is a prompt we can use to evaluate the output from an AI. Because the prompt we paste in includes the overarching goal, we don't need to restate it here—but it's worth keeping in mind that we need to evaluate the results in the context of that goal.

Please evaluate the below output from an AI. I would like you to evaluate the entire output according to the criteria in the attached Table. Use all of the criteria that apply, and use the same criteria to evaluate each of the examples in the AI output below. Please evaluate all of the entries in the AI output, using the same criteria in all cases. Please make suggestions about how to revise the output so that it is more accurate.

1. Goal: The output was in response to this prompt: [PASTE IN PROMPT HERE]
2. Criteria to be used for evaluation, in the attached Table. [ATTACH TABLE 3.1]
3. Output to be evaluated: [PASTE IN OUTPUT HERE]

And here is a sample of how GPT-4 responded:

The AI-generated output, consisting of six methods by which generative AI can assist in learning, will be evaluated based on the comprehensive criteria provided in Table 3–1. Here's a detailed evaluation of each entry using the applicable criteria:

1. Personalized Learning Experiences

- **Decomposing Complex Problems**: Successfully breaks down the concept of personalized learning into practical applications, benefits, drawbacks, and examples.
- **Vagueness and Ambiguity**: Mostly clear, but could further clarify what "basic access to technology" entails.
- **Inappropriate Assumptions**: Assumes minimal prior knowledge of AI without providing a rationale or addressing potential barriers for complete novices.
- **Evaluating Ethical Implications**: Appropriately considers data privacy and bias reinforcement.

- **Evaluating Evidence**: Relies on cited studies effectively, but could benefit from specifying how these studies control for variables.
- **Evaluating Causality and Relevance**: Discusses causality well but could better separate AI's role from other educational factors.
- **Trade-offs**: Well addressed, weighing scalability against costs and the need for bias monitoring.

Suggestions for Revision:

- Clarify what constitutes "basic access to technology."
- Justify the assumption of minimal prior AI knowledge or describe strategies to mitigate this barrier.
- Provide more details on the studies mentioned to ensure the claims meet the statistical significance criterion.

2. Simulation and Modeling Tools

- **Decomposing Complex Problems**: Effectively details the different aspects of using simulations.
- **Vagueness and Ambiguity**: Clearly explained; the term "moderate to high access to technology" could be quantified.
- **Evaluating Ethical Implications**: Addresses the risk of over-reliance well, but could expand on the ethical consider-ations of simulation accuracy.
- **Evaluating Evidence**: Could improve by discussing the stud-ies' methodologies in more detail.
- **Evaluating Causality and Relevance**: Adequately connects simulations with learning theory but could clarify causal rela-tionships better.
- **Trade-offs**: Identifies trade-offs between cost and benefits but might consider the long-term sustainability of technology use.

Suggestions for Revision:

- Define what "moderate to high access to technology" means in practical terms.
- Include more information about the computational require-ments and their long-term sustainability.

[TRUNCATED]

Without question, we can recruit AIs to help us think critically, which can in turn help us evaluate the results when we deploy the CAL.

Refining the Results

At this point, it's up to us to go through the results from the AI and decide—given the overall context of what we are doing—whether they achieve our goal. In some cases, we may want to follow up and ask the AI to elaborate on a point. For example, in this case I note that the AI did not consider "Inappropriate Assumptions" for "Simulation and Modeling Tools." If this omission undermines achieving the goal, we can easily go back and ask the AI to fill this in.

Based on such considerations, the next step of the CAL requires us to make a decision. As noted in the previous chapter, at this juncture we need to decide to take one of the following next steps:

1. The product is good enough to work with, and we thus we *retain* it. We now focus on tuning it up "by hand," such as by editing a document or touching up a graphic.
2. Although the product is close to what we want, it's not quite right, and so we *refine* the prompt. We edit it and then submit the new version, asking the AI to improve the previous results. I did this above, when I asked the AI to draw more on critical thinking criteria. As soon as the AI has responded to this revised prompt, we evaluate the new product and consider again whether it's good enough to retain and tune up or whether we need to continue to refine the prompt.
3. The product is so far from what we want that we *revise* the goal with a new one and return to Step 1.

All of the critical thinking that has come before sets us up for this decision. We need to decide which of these three options is most appropriate, given what critical thinking reveals about the limitations of the current results. In this particular example, I was happy enough with the results, given my goal.

After we decide which way to go, how do we actually implement the next step? In all three cases, we need more than just critical thinking. We now need to solve a problem: We need to figure out how to edit and tune up the results, refine the prompt, or revise the goal. In the following chapter we consider exactly how to address these sorts of problem-solving challenges.

Flourishing with the CAL

We need to learn to deploy a wide range of types of critical thinking if we are going to flourish in the Age of AI. As we've seen in this chapter, critical thinking is key to working well with AIs—and working well with AIs is crucial for gaining a sense of autonomy and control. Furthermore, working effectively and efficiently with AIs can help us perform better at work. This should not only contribute to feelings of job satisfaction, but also help us have stable jobs. Moreover, critical thinking informs our relationships with others and society writ large. Critical thinking is absolutely essential in an era of misinformation, disinformation, and increasingly wily bad actors.

But more than this, critical thinking contributes to the choices we make about personal growth, and the directions we take to develop our talents. Indeed, the first step in any decision—large or small—must be to analyze what the decision is about and what the choices are. We shall see that various forms of critical thinking are invoked in most of the activities we consider in the remainder of this book.

Critical thinking plays a special role in creative problem solving, and we often need to combine these two skills to flourish in the Age of AI. We consider exactly how to use the two skills together in the following chapter.

Digging Deeper

To dig deeper into the subject matter, below are examples of videos that provide a more in-depth treatment of aspects of the material reviewed in the chapter. In addition, the below search terms can help the reader locate new videos on the topic.

5 tips to improve your critical thinking—Samantha Agoos: https://www.youtube.com/watch?v=dItUGF8GdTw
Creativity and Critical Thinking in the Age of AI: https://www.youtube.com/watch?v=XhJN3gRCm8k

- "Critical thinking skills for the 21st century"
- "Developing critical thinking in the Age of AI"
- "Decomposing complex problems for critical analysis"
- "Identifying assumptions in arguments"
- "Evaluating ethical implications of decisions"
- "Assessing source credibility for critical thinking"
- "Evaluating evidence and relevance"
- "Logical consistency in arguments"

- "Distinguishing correlation and causation"
- "Decision making and critical thinking"
- "Feasibility analysis in critical thinking"

Using the Tools of Critical Thinking for Effective Decision Making: https://www.youtube.com/watch?v=REW3uV18YTw

The 5 Elements of Effective Thinking by Edward Burger & Michael Starbird: https://www.youtube.com/watch?v=2fSf-AmR9_A

- "Critical thinking in AI-assisted decision making"
- "Applying critical thinking to AI prompts"
- "Evaluating AI outputs with critical thinking"
- "Iterative critical thinking in AI interactions"
- "Critical thinking for refining AI-generated content"
- "AI and human collaboration in critical thinking"
- "Developing critical thinking skills with AI"
- "Ethical considerations in AI-assisted critical thinking"

The Intersection of Critical Thinking Skills and AI: Tactics for Effective Use: https://www.youtube.com/watch?v=7r1z-h-lfd4

Artificial Intelligence and Critical Thinking: https://www.youtube.com/playlist?list=PLqXQ0rOd491Kc4ElL8viVk-lfTrW1j41r

- "AI-assisted critical thinking skill development"
- "Using AI to learn critical thinking"
- "AI-generated scenarios for critical thinking practice"
- "Personalized critical thinking learning with AI"
- "Transfer of critical thinking skills from AI interactions"
- "Designing AI prompts for critical thinking development"
- "Evaluating critical thinking progress with AI"
- "Collaborative critical thinking learning with AI"

References

1   Scriven, M., & Paul, R. (1987). Critical thinking as defined by the National Council for Excellence in Critical Thinking, 1987. *8th Annual International Conference on Critical Thinking and Education Reform*. www.criticalthinking.org/pages/defining-critical-thinking/766

2   Barnett, S. M., & Ceci, S. J. (2002). When and where do we apply what we learn? A taxonomy for far transfer. *Psychological Bulletin, 128*, 612–637.

3   Haskell, R. E. (2004). Transfer of learning. In C. D. Spielberger (Ed.), *Encyclopedia of applied psychology*. Academic Press, pp. 575–586.

4   Hempenstall, K. (2019, September 25). *Near and far transfer in cognitive training*. National Institute for Direct Instruction. www.nifdi.org/resources/hempenstall-blog/758-near-and-far-transfer-in-cognitive-training.html

5   Herrity, J. (2024, April 9). How to write SMART goals in 5 steps (with examples). Indeed Career Guide. www.indeed.com/career-advice/career-development/how-to-write-smart-goals

6   Leonard, K., & Watts, R. (2024, July 9). The ultimate guide to SMART goals. *Forbes Advisor*. www.forbes.com/advisor/business/smart-goals/

Chapter 4

Elevating Creative Problem Solving

In July 2018, heavy rains flooded a cave in Thailand, and trapped the Wild Boars soccer team over a mile from the cave entrance. British divers discovered them nine days after they went missing, and reported back a particularly knotty problem: The cave was mostly flooded, the team members were weak, and most of the 12 boys, ages 11–15, couldn't swim.

Divers, rescue experts, and medical professionals from various countries around the world considered the problem from many angles. The rescue itself required cave divers, military personnel, doctors, and other volunteers to work together. The rescuers eventually decided to teach the boys very basic diving skills via remote instruction, and then to sedate them and move them individually through the flooded parts of the cave. This strategy had never been used before, and the rescuers had to continually monitor the health status of the boys, the water levels in the cave, and the weather—and they had to adapt their plans as the situation changed. They ultimately did succeed in saving the entire team.[1]

This rescue operation illustrates nicely the human ability to solve open-ended problems that require taking context into account. Many such examples exist.[2,3] As we use the term here, a "problem" occurs when we are confronted with a puzzle or challenge that we can only solve via original thinking. If we don't need to rely on original thinking, but instead just remember the solution from a previous encounter, it isn't really a "problem" for us. In some situations, the challenge is an obstacle or obstacles that must be overcome or circumvented, whereas in other situations the challenge is simply the need to go through a series of well-defined steps (e.g., as in common math problems) and discover whether they solve the problem, and in yet other situations, the challenge is to reach a goal with no clearly defined path towards it.

We humans are good at solving problems, including the sort of open-ended and context-dependent problems that are difficult for AIs (e.g.,[4,5,6]). However, we are still constrained by the limitations of our knowledge and of our brains—such as our limited ability to pay attention over sustained

DOI: 10.4324/9781032686653-5

amounts of time. Thus, even when we are faced with open-ended, context-dependent problems, we can benefit from using AIs as cognitive amplifiers.

We focus in this chapter on *creative* problem solving. This sort of problem solving relies on creative thinking aimed at achieving a particular goal. Not all problem solving requires creative thinking. We can solve some problems, such as common math problems, by following a set of well-worn steps. This is not creative thinking. In contrast, other problems require us to come up with something new, to be creative—such as was required to rescue the boys from the flooded cave. We focus here on this second type of problem.

Creative problem solving can occur in a very wide range of situations, such as when we are stuck in traffic and need to devise a new route, when we run out of ink in our printer but need to make a hard copy, or when we have to devise a new business plan. Creative problem solving relies on creative thinking, which is thinking that produces something novel and useful. As we shall see, we can deploy the Cognitive Amplifier Loop (CAL) to help us engage in creative problem solving.

Like critical thinking, creative problem solving draws on a set of distinct techniques. These techniques are employed in two phases: divergent and convergent thinking.[7,8,9] *Divergent thinking* requires us to generate many ideas for candidate solutions to a problem, whereas *convergent thinking* requires us to refine and narrow down these ideas to find the most promising one. Divergent thinking comes first, followed by convergent thinking, but a round of convergent thinking may lead to another round of divergent thinking, and so forth. The process of creative problem solving is neither linear nor lockstep.

Divergent Thinking

When we engage in divergent thinking, we should be spontaneous and focus on both novelty and quantity. We should not consider how good the candidate solutions are. Even bad ideas can be useful, for example by triggering a good idea or by later being combined with a good idea to produce something even better. Thus, we shouldn't worry about whether the ideas are any good at this point—we just produce as many new ideas for solutions as we can. Once we have a slew of candidate solutions, we then will use a lot of critical thinking to winnow them down—but we should hold off on this process until we have generated as many novel candidate solutions as possible.

To illustrate, let's suppose that we are stuck on how to rescue those boys from the flooded mine described at the outset of this chapter. Perhaps we knew that we needed to send people in to help the children who couldn't swim, but we weren't exactly sure how this could work. We know that we will recognize what we want when we see it, but aren't sure how to produce candidate solutions. So, now what?

The SCAMPER and AFREETA Methods

The SCAMPER method is a tried-and-true way to develop alternative ideas, which we can employ to produce a set of possible solutions to a problem.[10,11,12] The method is a collection of seven techniques; the acronym stands for Substitute, Combine, Adapt, Modify, Put to another use, Eliminate, and Rearrange. Table 4.1 summarizes the method in the context of devising new solutions, with examples of different possible solutions to the problem of rescuing people trapped in a flooded mine. In this table, I treated one candidate solution as the goal; in the first four rows of Table 4.1, I could have just as easily replaced "goal" with "candidate solution."

We need to do a lot of work to employ SCAMPER. To get a sense of why using this method is laborious, let's illustrate with the very first technique, Substitution. This technique requires us to think about what we can use instead of one or more existing elements. To deploy this technique in divergent thinking, we engage in the following steps:

First, we need to start by identifying the components of the candidate solution. By rights, this is part of critical thinking—but critical thinking and creative thinking are intimately connected, and this is indeed an essential first step in divergent thinking. Second, we now evaluate each component

Table 4.1 The SCAMPER Method.

Method	Description	Example
Substitute	Can we substitute a new element for part of the previous goal?	Using a drone to deliver supplies instead of sending divers.
Combine	Can we combine parts of the original goal?	Combining the use of pumps and sealing leaks to manage water levels.
Adapt	Can we adapt a different goal to produce a new one?	Adapting scuba diving rescue techniques for mine conditions.
Modify	Can we modify the original goal to produce a new one?	Modifying the equipment to withstand higher water pressure.
Put to another use	Can we put to another use an existing concept?	Using mining equipment to drill additional escape routes.
Eliminate	Can we eliminate a feature or constraints?	Eliminating non-essential rescue equipment to speed up operations.
Rearrange	Can we rearrange components, perspectives, or roles?	Rearranging the rescue team shifts to maintain round-the-clock operations.

for its importance, function, and potential for substitution. We need to keep the general goal in mind here; there's no point in veering off-track, so that a candidate solution is irrelevant or inappropriate. And third, we need to research alternatives. We look for methods, concepts, objects, or anything else that can serve as potential substitutes.

As should be evident, this process is laborious and often tedious. Such situations cry out for the CAL.

Indeed, the situation is worse than it may appear because the SCAMPER method is not an exhaustive set of techniques. In particular, based on empirical studies and best practices, it's worth deploying seven additional techniques, subsumed by the acronym AFREETA: Analogies, Forcing connections, Reframing obstacles, Edge cases, Elaborating associations, Trade opposites, and Adding constraints. Table 4.2 summarizes these techniques.

Table 4.2 The AFREETA Method.

Method	Description	Example
Analogies	Can we use an analogy to create new possible solutions?	Using the analogy of ants creating tunnels to inspire new ways to drill escape routes.
Forcing connections	What connections can we force?	Connecting existing water pumps with modified air pumps to create a dual-purpose system.
Reframing obstacles	Can we reframe obstacles to make them an opportunity?	Viewing the water as a transportation medium rather than a barrier, and using boats or flotation devices.
Edge cases	Can we create new goals by imagining scenarios under unusual or extreme conditions or focus on rare or atypical cases that fall outside of the norm?	Considering the scenario where the cave is completely submerged and planning for underwater escape routes.
Elaborating associations	Given associations that already exist, how can we expand on them?	Expanding the use of communication lines to also deliver supplies and instructions.
Trade opposites	Can we reverse meanings of terms and produce something useful?	Instead of focusing on keeping water out, find ways to manage and utilize the incoming water.
Adding constraints	Can we create new goals by adding constraints, such as by imposing limitations on resources?	Imposing a time constraint to force the team to develop quicker and more efficient rescue methods.

Again, this is a lot of work. Consider just the first of these techniques, where we ask *Can we use an analogy to create new possible solutions?* An analogy requires us to use characteristics of one object, situation, or phenomenon to explain another. Analogies can facilitate creative problem solving because they allow us to take known solutions or concepts from one domain and apply them to another domain, often leading to breakthrough ideas.

To use the analogy technique, we engage in the same three steps we employed with the SCAMPER techniques. First, we must utilize critical thinking to analyze the components, but now we need to pay particular attention to the relations among those components—which are crucial for any analogy.[13,14] For example, if we were trying to figure out how to combat a disease that was killing plants, we might think about the components of a plant cell and how they relate to each other. We would identify the cell wall, which provides structure and protection; the chloroplasts, which produce energy through photosynthesis; the nucleus, which controls cell activities; and the vacuole, which stores nutrients and waste products. Second, we would identify potential parallel domains. For example, we might choose a city, which can have city walls or boundaries, power plants, a city hall or government, and waste management and storage facilities. And third, we specify the relationships, noting the point-for-point correspondences between the two domains, such as cell wall / city walls or boundaries, chloroplasts / power plants, nucleus / city hall or government, and vacuole / waste management and storage facilities. This analogy might lead us to think about how firefighters and police work together during a fire, and then to consider possible corresponding functions in cells that could be marshaled to fight the disease.

We've now considered 14 different techniques for divergent thinking. Depending on the problem at hand, some are more relevant than others. But even so, there are too many techniques to keep in mind and too many to try out.

When faced with human limitations, such as our inability to hold many things in mind simultaneously, we should immediately think of AIs as cognitive amplifiers. Here's a prompt to engage an AI in divergent thinking about another example of a rescue problem, followed by how it responded.

Me:

Please create possible solutions to the problem noted below. To do so, use the techniques summarized in the two attached tables, which describe the SCAMPER and the AFREETA techniques.

Problem: A ferocious storm has been raging for a week, and has trapped a ski team at the top of a remote mountain. The team has no shelter, and has been hiding behind some boulders, which do not provide much protection. They also have no food. Some of them have become so weak that they cannot walk. The storm is projected to last at least another three days, which will be too late for some of the team members.

Tables: The two tables are attached. [ATTACH TABLES 4.1 AND 4.2]

AI:

Solutions Using the SCAMPER Method

1. Substitute:

   - Use drones to deliver supplies: Instead of waiting for rescue teams to reach the ski team, use drones to deliver essential supplies such as food, thermal blankets, and medical kits to the trapped team members. This would provide immediate relief and sustain them until the storm passes.

2. Combine:

   - Combine shelter and heat sources: Use available materials (e.g., ski equipment, boulders) to create makeshift shelters that can also trap and retain heat. Additionally, use portable stoves or heating pads that could be delivered by drone.

3. Adapt:

   - Adapt survival techniques from other environments: Apply survival techniques from desert or jungle environments, such as creating windbreaks or snow caves, to provide better shelter and protection from the storm.

4. Modify:

   - Modify equipment for better use: Alter skiing equipment (like poles and skis) to build a more robust shelter or to create signals for rescuers.

5. Put to another use:

   • Use skiing gear for different purposes: Use ski poles as markers for rescue teams or skis as digging tools to create a snow cave for better shelter.

6. Eliminate:

   • Eliminate unnecessary equipment: Focus on essential survival gear. Get rid of any non-essential items to make room for more critical supplies like food and thermal blankets that can be delivered via drones.

7. Rearrange:

   • Rearrange the team roles: Assign roles based on each member's current physical condition to optimize the use of their remaining energy and skills, such as those still strong enough creating better shelters or signaling for help.

Solutions Using the AFREETA Method

1. Analogies:

   • Use the analogy of huddling animals: Like how animals huddle together for warmth, the team can use body heat and shared insulation techniques to maintain body temperature.

2. Forcing connections:

   • Combine existing technologies: Use emergency flares or reflective materials on the skis to create signals that are more visible to rescuers.

3. Reframing obstacles:

   • View the storm as a resource: Use the snow from the storm to build more substantial shelters like snow caves, which can provide better insulation against the cold.

4. Edge cases:

   • Imagine extreme scenarios: Plan for the possibility that the storm might last longer. Consider rationing food more strictly and preparing mentally and physically for prolonged survival.

5. Elaborating associations:

   • Expand on existing communication methods: Use mobile phones or emergency radios to communicate with rescuers. If battery life is an issue, consider using hand-crank chargers or solar chargers.

6. Trade opposites:

   • Manage the cold instead of avoiding it: Focus on techniques that leverage body heat and any available insulation to manage the extreme cold instead of trying to eliminate it.

7. Adding constraints:

   • Impose a strict timeline for creating shelters: Set a strict time-line for building more robust shelters, which forces the team to act quickly and efficiently before their energy levels drop further.

By applying the SCAMPER and AFREETA techniques, we can generate a variety of innovative solutions to help the trapped ski team survive the storm and increase their chances of being rescued.

Some of these suggestions are clearly better than others, but it's striking how many are worth considering. We could just as easily ask the AI to give us five examples produced with each technique. Clearly, the CAL can help us in divergent thinking.

Convergent Thinking

Once have deployed the CAL to generate many candidate solutions for a problem, we then need to select among them. Whereas divergent thinking opens up the space of possibilities, convergent thinking narrows down these options. Convergent thinking is systematic and logical. As such, convergent thinking invokes critical thinking to evaluate the possible solutions and identify the most promising one. We can put to good use many of the types of critical thinking we reviewed in Chapter 3. In addition, however, researchers and practitioners have developed specific techniques for just this purpose. Once we consider key examples of such techniques, we can see how we can utilize an AI to help us sort among the possibilities, just as we can use one to help us produce them in the first place.

Three specific techniques can be especially useful in helping us sort among candidate solutions, in addition to the general critical thinking techniques summarized in Table 3.1. Namely, we can ask, *What are the relative strengths, weaknesses, opportunities, and threats? What are the costs vs. the benefits? Which biases are at work, what is their impact, and how can we counter them?*

SWOT Analysis

First, consider a "SWOT analysis," which assesses the *Strengths, Weaknesses, Opportunities*, and *Threats* of any proposed course of action (e.g.,[15,16]). To illustrate the method, let's consider just two solutions to the ski team rescue problem we just considered, and see how a SWOT analysis can help us decide which one to prioritize.

Me:

Please help me use a SWOT analysis to evaluate two solutions to this problem: Problem: A ferocious storm has been raging for a week, and has trapped a ski team at the top of a remote mountain. The team has no shelter, and has been hiding behind some boulders, which do not provide much protection. They also have no food. Some of them have become so weak that they cannot walk. The storm is projected to last at least another three days, which will be too late for some of the team members. The solutions are as follows:

Substitute: Use drones to deliver supplies: Instead of waiting for rescue teams to reach the ski team, use drones to deliver essential supplies such as food, thermal blankets, and medical kits to the trapped team members. This would provide immediate relief and sustain them until the storm passes.

Combine: Combine shelter and heat sources: Use available materials (e.g., ski equipment, boulders) to create makeshift shelters that can also trap and retain heat. Additionally, use portable stoves or heating pads that could be delivered by drone.

AI:

SWOT Analysis of Two Solutions for the Ski Team Trapped on a Mountain

1. Substitute: Use drones to deliver supplies

Strengths

- Immediate Delivery: Drones can provide quick delivery of essential supplies such as food, thermal blankets, and medical kits, which are critical for survival.
- Accessibility: Drones can reach areas that are otherwise inaccessible due to the storm or rugged terrain.
- Reduce Risk: Using drones minimizes the risk to human rescuers who would otherwise have to navigate dangerous conditions.

Weaknesses

- Dependency on Weather: Drone operations can be heavily impacted by severe weather conditions, which might make flights risky or impossible.
- Limited Payload: Drones can only carry a limited amount of supplies, which may not be sufficient for the entire team or for an extended period.
- Technical Failures: Potential for malfunctions or loss of drones due to harsh conditions or navigational errors.

Opportunities

- Innovation in Rescue Operations: Successful use of drones can pave the way for their future use in similar rescue operations, setting a precedent for high-tech rescue solutions.
- Partnerships: Opportunities to collaborate with tech companies specializing in drone delivery services, which could enhance the capabilities and reliability of the rescue efforts.

Threats

- Regulatory Restrictions: There might be airspace restrictions or regulatory hurdles for flying drones in certain areas, especially near mountainous regions.
- Public Perception: Failure in the mission could lead to negative publicity regarding the reliability of using drones for critical rescue operations.

2. Combine: Combine shelter and heat sources

Strengths

- Use of Available Resources: Utilizes readily available materials like ski equipment and boulders to create makeshift shelters, which is cost-effective and immediate.
- Improved Protection: Shelters can trap and retain heat, crucial for surviving harsh conditions and preventing hypothermia among the team members.

Weaknesses

- Labor Intensive: Requires physical effort to construct shelters, which could be challenging given the weakened state of some team members.
- Sustainability: The effectiveness and durability of the shelters could be limited, especially if the storm continues to worsen.
- Dependence on Additional Supplies: The success of this solution might still rely on the delivery of external supplies like heating pads, which could be impacted by the same limitations affecting drone deliveries.

Opportunities

- Skill Development: Team members can develop or utilize survival skills, which could be beneficial in both the current situation and future endeavors.
- Team Cohesion: Working together to build shelters can improve morale and foster a sense of unity and purpose among the team members.

Threats

- Insufficient Shelter: There is a risk that the shelters built may not provide adequate protection if the weather deteriorates further, leading to potential health risks.
- Resource Limitations: There may be a limited supply of materials that can be used for shelter, and inappropriate or inadequate shelter construction could expose the team to greater risks.

Conclusion
Both solutions have their unique strengths and could potentially be combined for a more comprehensive approach to the problem. The use of drones offers a quick and risk-free method to deliver essential supplies, while combining shelter and heat sources provides a sustainable way to enhance the team's chances of surviving the harsh conditions until rescue can arrive. Each solution also faces significant challenges that need to be carefully managed to ensure the safety and well-being of the trapped ski team.

This analysis can lead to insights, but how, exactly, should we act on it? One way to do this is to assign each aspect of the SWOT analysis a numerical score. We can rate Strengths and Opportunities from 0–5, and we can rate Weaknesses and Threats from 0 to -5. We can then sum up the four ratings, and get a sense of the overall utility of each proposed solution. For example, in the above example, for the Drone solution we might rate the Strengths as 4 and Opportunities as 4 whereas we might rank the Weaknesses as -3 and the Threats as -1. The total would then be 4. In contrast, for the Shelter and Heat Source solution, we might rate the Strengths as 3 and Opportunities as 4 whereas we might rank the Weaknesses as -4 and the Threats as -4. The total would then be -1. Comparing the two, we would favor the Drone solution.

Ideally, we should have clear criteria for making these ratings, and should conduct this sort of analysis for every candidate solution that comes out of the divergent thinking phase. This might work if we have only a few such candidates, but would quickly get out of hand if we have many candidates. This problem calls out for the CAL. We could easily create a rubric that defines clear criteria for evaluating the solutions, and prompt an AI to draw on it when making its judgments (cf. [17]).

Cost–Benefit Analysis

Another way to assess trade-offs among options is a *cost–benefit analysis*, which lays bare the economic value of an idea by comparing the expected costs of implementing it to the anticipated benefits.[18,19] Again, this is a good use of the CAL, employing an AI to relieve a cognitive burden and allow us to focus on what we do particularly well—in this case, putting the proposals in context. I simply asked GPT-4 to do a cost–benefit analysis of the two solutions we just considered, and here's what it said:

Cost–Benefit Analysis of Two Solutions for the Ski Team Trapped on a Mountain

1. Substitute: Use drones to deliver supplies

Benefits

- Rapid Response: Quick delivery of essential survival supplies like food, medical kits, and thermal blankets.
- Safety: Minimizes the risk to human rescuers by using unmanned vehicles in dangerous conditions.
- Versatility: Drones can potentially perform multiple flights, delivering various types of aid as needed.

Costs

- Financial: High cost associated with deploying drones, including technology, operation, and possible rental or purchase costs.
- Reliability Risks: Potential for drone failure due to harsh weather, leading to financial losses and unmet needs.
- Regulatory and Coordination Costs: Possible costs related to obtaining necessary clearances and coordinating with multiple agencies.

2. Combine: Combine shelter and heat sources

Benefits

- Immediate Shelter: Provides immediate protection using available resources, improving the team's ability to survive.
- Heat Retention: Enhances the ability to maintain body temperature, crucial under severe weather conditions.
- Psychological Boost: Building and improving the shelter can offer psychological benefits by keeping the team members active and focused.

Costs

- Physical Effort: Significant energy and effort required, which might be taxing for already weakened team members.

- Dependence on External Factors: Relies on the successful delivery of additional supplies like heating pads, which could be hindered by the same issues affecting drone deliveries.
- Temporary Solution: The shelters may provide only temporary relief, possibly requiring additional efforts and resources if the situation persists.

Cost–Benefit Evaluation
Use drones to deliver supplies

- Effectiveness: High, provided the drones can operate in the given weather conditions.
- Cost Efficiency: Medium to high, depending on the drone technology available and the scale of deployment. While upfront costs are high, the potential to save lives and minimize traditional rescue risks can justify the expense.
- Reliability: Moderate to low, heavily dependent on weather conditions and technical capabilities.

Combine shelter and heat sources

- Effectiveness: Moderate to high, heavily dependent on the ability and energy levels of team members to construct and maintain shelters.
- Cost Efficiency: High, as it primarily uses existing resources and manpower, with additional costs for supplemental heating elements if delivered.
- Reliability: Moderate, as the sustainability of shelters may vary based on weather changes and material availability.

Conclusion
While both solutions have notable strengths, they also present significant risks and costs. The use of drones, despite higher financial costs, offers a non-invasive and rapid delivery method that could be crucial in urgent situations. However, its success is highly dependent on external factors like weather and regulatory permissions. Combining shelter and heat sources provides a cost-effective solution but may not suffice alone, especially under deteriorating

conditions. A combined approach that integrates both strategies might offer a more robust solution, balancing the immediate benefits of drone deliveries with the sustainable shelter improvements provided by the second solution. This hybrid approach could potentially maximize benefits while mitigating the individual drawbacks of each solution.

It's interesting to see that the AI came up with a hybrid approach on its own, trying to get the best of both solutions. Cost–benefit analyses are particularly effective when we can assign numbers to the various costs and benefits. But, as in the problem considered here, in many cases some of the values are imponderable. For example, we cannot assign a particular dollar value to saving a human life. Nevertheless, even this qualitative approach is useful. In general, we need to treat these AI outputs as suggestions, which we then revise as we see fit. They clearly are useful as ways to prime the creative pump!

Identifying and Combating Bias

An additional technique can help us in the convergent thinking process. We now ask, *Which biases are at work, what is their impact, and how can we counter them?*

Many types of bias can affect our decisions.[20,21,22,23] However, this is not always bad. For example, if an alternative solution evokes positive emotions, this can be a good thing. Nevertheless, we should be aware of these biases and how they are affecting us, and be prepared to counter them if they are distorting our judgment. Table 4.3 summarizes biases that are likely to affect the convergent phase of creative problem solving, as well as common ways to try to counter the effects of these biases. Imagine trying to hold all of this in mind!

Note that different biases in Table 4.3 may tug us in different directions. For example, some biases—such as Confirmation Bias, Optimism/Pessimism, Sunk Cost Fallacy, and the Escalation of Commitment—may tend to lead us to see what we want to see. In contrast, other biases—such as the Status Quo and Bandwagon Bias—may make us too sensitive to what others think and do.

The sheer complexity of the range and types of biases invites us to use an AI as a cognitive amplifier. An AI can help us identify and navigate around these biases. However, we need to be careful to eliminate biases when writing the prompt. For example, we could inadvertently fall prey to

Table 4.3 Biases That Can Affect Convergent Thinking.

Bias	Description	Mitigation	Example
Confirmation Bias	Seeking, interpreting, and/or remembering information that confirms our preconceptions.	Actively seek out and consider evidence that contradicts our beliefs.	Assuming the storm will clear up soon based on optimistic weather reports, ignoring more conservative forecasts.
Anchoring Bias	Relying too heavily on the first piece of information encountered when making decisions.	Gather more information before making a decision and think critically about initial information.	Overemphasizing initial reports of a three-day storm extension, ignoring updates indicating it may last longer.
Availability Heuristic	Overestimating the importance of information that is readily available or recent.	Look for objective data and statistics rather than relying solely on personal experience or anecdotal evidence.	Prioritizing recent successful rescues from similar situations without considering unique challenges of the current scenario.
Overconfidence Bias	Being more confident in our abilities or opinions than is objectively justified.	Seek feedback, consider different viewpoints, and review past decision outcomes to calibrate confidence levels.	Overestimating the team's ability to survive without external help, despite worsening conditions.
Sunk Cost Fallacy	Continuing a course of action because of past investments (time, money, resources) regardless of current prospects.	Make decisions based on future benefits and costs, not past investments.	Persisting with waiting for traditional rescue teams despite evidence that drones could deliver supplies faster.

(Continued)

Table 4.3 (Continued)

Bias	Description	Mitigation	Example
Status Quo Bias	Preferring things to stay the same by doing nothing or by sticking with a decision made previously.	Regularly review and question existing choices and be open to change when warranted.	Sticking with initial survival strategies despite the availability of better options as conditions change.
Groupthink	Conforming to the prevailing opinions within the group, leading to a decrease in critical evaluation.	Encourage open dialogue, dissent, and independent thinking within groups.	The team unanimously decides to wait for rescue without considering alternative survival tactics.
Framing Effect	Accepting decisions based on how information is presented rather than based on the information itself.	Try to reframe the information in various ways to see whether or how the framing affects your decision.	Being swayed against using drone deliveries because they are framed as "experimental" rather than "innovative."
Hindsight Bias	Believing after an event has occurred that we would have predicted or expected the outcome.	Keep records of predictions and compare them with actual outcomes.	Believing that the severity of the storm was obvious from the start, leading to overconfidence in future predictions.
Dunning-Kruger Effect	Noting that the less knowledgeable or skilled individuals are, the more likely they are to overestimate their abilities.	Pursue continuous learning and self-improvement and seek feedback from competent sources.	Team members with limited survival training assert they can manage the situation without external help.
Optimism/ Pessimism Bias	Being overly optimistic or pessimistic about outcomes.	Seek balanced perspectives and consider both positive and negative aspects equally.	Being overly pessimistic about rescue prospects, leading to inaction, or overly optimistic, leading to underpreparedness.

Table 4.3 (Continued)

Bias	Description	Mitigation	Example
Bandwagon Effect	Doing or believing things because many other people do or believe the same.	Make decisions based on evidence and analysis rather than following the crowd.	Deciding to use certain survival techniques because they are popular, without considering their applicability to the current situation.
Choice-Supportive Bias	Remembering our choices as better than they actually were.	Keep an objective record of outcomes and regularly review decisions to assess their actual effectiveness.	Recalling initial survival strategies as more effective than they were, leading to resistance to change tactics.
Fundamental Attribution Error	Attributing others' actions to their character or personality while attributing our own actions to situational factors.	Consider situational factors that might influence others' behaviors and avoid jumping to conclusions about their motives.	Attributing the delay in rescue efforts to incompetence, ignoring the extreme weather conditions hampering rescue operations.
Projection Bias	Assuming that others share our beliefs, values, or preferences.	Actively seek to understand others' perspectives and recognize that their views and experiences may be different.	Projecting personal survival strategies onto the team, assuming everyone has the same capabilities and preferences.
Escalation of Commitment	Increasing commitment to a decision in spite of negative information or outcomes, often to justify previous commitments.	Regularly reassess ongoing projects or decisions in light of new information and be willing to change course if necessary.	Continuing to wait for rescue despite worsening conditions and evidence that immediate action is necessary.

the Optimism/Pessimism bias in the way we write the prompt. Thus, after we draft a prompt, we can submit it to an AI and ask whether the prompt embodies any of the biases noted above. The prompt can be as simple as the following:

Below is a prompt that I want to submit to an AI. I would like you to use information in the attached Table to evaluate whether this prompt embodies any of the biases summarized there. If it does, could you please revise the prompt to eliminate this problem?

Prompt: [INSERT PROMPT HERE]
Table: [ATTACH TABLE 4.3]

Using the CAL to Converge on a Solution

Sometimes we aren't sure which handful of candidate solutions might be most promising, and so want to rank them all—not just compare a few top contenders. In this case we can prompt an AI to provide such rankings. To continue with the example of rescuing the ski team trapped at the top of the mountain, we can deploy a prompt like the following to help us rank order the alternative solutions we created using the SCAMPER and AFREETA techniques. In order to provide pertinent background information, I pasted Tables 3.1 and 4.3 into Word documents, which I then uploaded to the AI when I submitted the prompt below. To clearly illustrate this approach, I had the AI use only the criteria in these two tables, but we could have had it also employ the specialized techniques. The AI's responses follow the prompt:

Me:

Please help me rank-order possible solutions to a problem, with a rank of "1" indicating the best, a rank of "2" indicating the second best, and so on. Please rank all of the potential solutions, from beginning to end. Part 1 below describes the precise task. Part 2 describes the nature of the problem to be solved. Part 3 provides the various candidate solutions that I would like you to rank. Part 4 provides two Tables.

Part 1: Please use the critical thinking techniques summarized in the attached Table 3.1 to rank order the solutions, from beginning to end. For each individual solution, please: a) explain the reasoning behind your ranking, and b) use Table 4.3 to flag potential biases that may have affected the ranking. Please note all potential biases that are relevant, not just the top two or three.

Part 2. The problem: Problem: A ferocious storm has been raging for a week, and has trapped a ski team at the top of a remote mountain. The team has no shelter, and has been hiding behind some boulders, which do not provide much protection. They also have no food. Some of them have become so weak that they cannot walk. The storm is projected to last at least another three days, which will be too late for some of the team members.

Part 3: Alternative candidate solutions:

- Use drones to deliver supplies: Instead of waiting for rescue teams to reach the ski team, use drones to deliver essential supplies such as food, thermal blankets, and medical kits to the trapped team members. This would provide immediate relief and sustain them until the storm passes.
- Combine shelter and heat sources: Use available materials (e.g., ski equipment, boulders) to create makeshift shelters that can also trap and retain heat. Additionally, use portable stoves or heating pads that could be delivered by drone.
- Adapt survival techniques from other environments: Apply survival techniques from desert or jungle environments, such as creating windbreaks or snow caves, to provide better shelter and protection from the storm.
- Modify equipment for better use: Alter skiing equipment (like poles and skis) to build a more robust shelter or to create signals for rescuers.
- Use skiing gear for different purposes: Use ski poles as markers for rescue teams or skis as digging tools to create a snow cave for better shelter.
- Eliminate unnecessary equipment: Focus on essential survival gear. Get rid of any non-essential items to make room for more critical supplies like food and thermal blankets that can be delivered via drones.

- Rearrange the team roles: Assign roles based on each member's current physical condition to optimize the use of their remaining energy and skills, such as those still strong enough creating better shelters or signaling for help.
- Use the analogy of huddling animals: Like how animals huddle together for warmth, the team can use body heat and shared insulation techniques to maintain body temperature.
- Combine existing technologies: Use emergency flares or reflective materials on the skis to create signals that are more visible to rescuers.
- View the storm as a resource: Use the snow from the storm to build more substantial shelters like snow caves, which can provide better insulation against the cold.
- Imagine extreme scenarios: Plan for the possibility that the storm might last longer. Consider rationing food more strictly and preparing mentally and physically for prolonged survival.
- Expand on existing communication methods: Use mobile phones or emergency radios to communicate with rescuers. If battery life is an issue, consider using hand-crank chargers or solar chargers.
- Manage the cold instead of avoiding it: Focus on techniques that leverage body heat and any available insulation to manage the extreme cold instead of trying to eliminate it.
- Impose a strict timeline for creating shelters: Set a strict timeline for building more robust shelters, which forces the team to act quickly and efficiently before their energy levels drop further.

Part 4: Attached please find Table 3.1, which summarizes types of critical thinking to use when ranking the possible solutions, and Table 4.3, which summarizes types of bias that may have affected each ranking. [ATTACH TABLES 3.1 AND 4.3]

AI:

Comprehensive Rank-Order of Solutions for the Ski Team Rescue Scenario

Part 1: Solution Analysis

1. Use drones to deliver supplies

   - Reasoning: This offers immediate relief by providing food, medical supplies, and warmth. It's a direct response to the urgent needs for survival.
   - Biases:
   - Optimism Bias: Might overestimate drone capability in bad weather.
   - Overconfidence Bias: Assuming no technical failures with drones.
   - Availability Heuristic: Recent successful drone operations may bias the perception of their reliability in extreme conditions.

2. Combine shelter and heat sources

   - Reasoning: Creates immediate, sustainable shelter using on-hand resources and possibly delivered materials, improving the team's exposure resistance.
   - Biases:
   - Anchoring Bias: Initial success may prevent considering alternative shelter options.
   - Confirmation Bias: Favoring familiar methods without assessing their full suitability in the current extreme conditions.

3. Rearrange the team roles

   - Reasoning: Utilizes the current capabilities of the team members efficiently, ensuring those who are stronger help those in weaker conditions.
   - Biases:
   - Groupthink: Might lead to unchallenged poor role assignments.
   - Projection Bias: Assuming all team members will function well in assigned roles.

4. Use the analogy of huddling animals

   - Reasoning: Promotes shared body heat retention, simple and effective in immediate term.
   - Biases:
   - Framing Effect: Positively framing this method may overlook its potential limitations, like individual vulnerabilities to cold.

5. View the storm as a resource

   - Reasoning: Uses snow for constructing substantial shelters, which can provide significant insulation.
   - Biases:
   - Optimism/Pessimism Bias: Could be overly optimistic about the practicality and effectiveness of building with snow under current conditions.

6. Modify equipment for better use

   - Reasoning: Adapts available resources for multiple survival needs, potentially increasing chances of being located or improving shelter.
   - Biases:
   - Status Quo Bias: Might limit creative uses of equipment if sticking too closely to their conventional uses.

7. Use skiing gear for different purposes

   - Reasoning: Similar to modifying equipment but specifically focuses on using skis and poles for practical survival tasks.
   - Biases:
   - Escalation of Commitment: Continuing to use these items in traditional ways despite evidence suggesting other uses could be more beneficial.

8. Combine existing technologies

   - Reasoning: Uses flares and reflective materials to enhance rescue visibility, which could speed up rescue operations.
   - Biases:
   - Bandwagon Effect: Following common rescue practices without adapting them to the specific situation might reduce effectiveness.

9. Expand on existing communication methods

   - Reasoning: Maintains open lines of communication with rescuers, which is crucial for coordination.
   - Biases:

- Hindsight Bias: Overestimating the effectiveness of communication based on past successes.

10. Eliminate unnecessary equipment

   - Reasoning: Focuses on essential survival items, potentially making movement and management easier for weakened team members.
   - Biases:
   - Sunk Cost Fallacy: Reluctance to discard items that might have cost but are not useful currently.

11. Manage the cold instead of avoiding it

   - Reasoning: Focuses on using body heat and available materials to cope with the cold, a more realistic approach given the circumstances.
   - Biases:
   - Fundamental Attribution Error: Misjudging the team's ability to effectively manage the cold based on their current physical state.

12. Impose a strict timeline for creating shelters

   - Reasoning: Encourages prompt action to maximize energy use before conditions worsen.
   - Biases:
   - Overconfidence Bias: Assuming that setting strict deadlines will automatically lead to efficient work under extreme stress and fatigue.

13. Imagine extreme scenarios

   - Reasoning: Prepares for worst-case scenarios, potentially extending survival time.
   - Biases:
   - Dunning-Kruger Effect: Overestimating ability to handle extended scenarios without adequate preparation or resources.

Part 2: Application of Critical Thinking Techniques from Table 3.1

- Define the Problem: Directly addressing the critical needs of shelter, warmth, and nutrition.
- Questioning Assumptions: Evaluating the effectiveness of traditional methods under unusual circumstances.
- Evaluate Ethical Implications: Balancing immediate needs against potential long-term survival.
- Evaluate Data and Claims: Assessing the reliability of technologies like drones in extreme weather.
- Analyze Inferences: Ensuring logical consistency in survival strategy planning.
- Make Decisions: Weighing pros and cons, considering both the technical feasibility and operational practicality of each solution.

By following this comprehensive approach and being mindful of biases, a more objective and practical solution order can be achieved, significantly improving the chances of the ski team's survival.

The ranking strikes me as reasonable. Notice, however, that the AI stopped providing multiple biases after the third candidate. We could easily follow up with a second prompt, asking it to flesh out the additional biases that could have affected the solutions ranked under the top three. However, even with this oddity, the output vastly simplifies the problem, and sets us up to do what we do well. As noted previously, it's much easier for us humans to recognize something than to recall it, and this approach plays to that strength.

The Human Touch

After we have decided to accept a particular result from the AI, we probably want to edit it to achieve our goal as closely as possible. We edit a result for clarity, style, and effect. We can treat each of these criteria as the subject of another round of problem solving. In our example of proposing a method for rescuing the ski team, we could use an AI to help us write a report or produce a post-operation PowerPoint presentation. We can have an AI produce the first draft, but then we need to edit it, as follows.

Edit for Clarity

To edit for clarity we need to know what to add, what to eliminate, and what to modify. This sort of problem solving is a good example of the interplay between critical thinking and creative thinking. We need to return to the goal and compare what we obtained to what we sought.

When editing for clarity, we need to put ourselves in the shoes of our intended audience, and anticipate how they would receive the product. To do so, we can use our intuitive understanding of "Theory of Mind" to guide us in revising the final product from the CAL. Theory of Mind is the cognitive ability that allows us to infer other people's mental states. Researchers have proposed two classes of theories to explain this ability. The "Simulation Theory" posits that we infer others' mental states by simulating or imagining ourselves in that other person's situation. According to this view, we use our own mental processes as a model to infer what others are thinking or feeling. The "Theory Theory," in contrast, states that we use implicit theories about how minds work to infer others' mental states. According to this type of theory, we infer another person's beliefs, desires, and intentions much like a scientist who forms a hypothesis to explain observations. According to this theory, people develop a set of cognitive rules and principles about human behavior and draw on this knowledge to interpret others' actions.

Studies have provided support for both classes of theories. For example, Simulation Theory is supported by the finding that the same parts of the brain are activated when we register other people's behavior and when we initiate that behavior ourselves.[24] Theory Theory is supported by the finding that children develop Theory of Mind in tandem with their development of more general reasoning abilities, which may suggest that reasoning underlies Theory of Mind.[25] But this is a correlation, which does not prove causation. Moreover, we must be mindful of the fallacy of the excluded middle (see Chapter 3), and realize that both mechanisms might operate, either at the same time or in different situations. In fact, this integrative approach is supported by neuroscientific studies showing that different brain networks may be involved in different aspects of Theory of Mind, possibly correlating with the mechanisms proposed by Simulation Theory and by Theory Theory.[26]

Hence, when editing a product that we want to show to others, we need to "run a mental simulation" and also reason about how they are likely to receive it, and strive to highlight the most important aspects and eliminate potential ambiguity.

Edit for Style

Different styles—in both language and design—are appropriate for different purposes and audiences. For example, we would adjust how formal the tone is, depending on the audience and context. Theory of Mind helps us infer the proper tone, and hence also figures into the style we adopt. In addition, if we are designing products, we need to ensure that the product's design, such as packaging or the product interface, aligns with the brand's aesthetic, using consistent fonts, colors, and design motifs that reflect the brand's style. Similarly, if we include an illustration, it must be in a style that complements the message. For example, it might be minimalist and clean for tech gadgets vs. warm and fuzzy for handmade goods. In addition, we need to align the visual style of an app with design principles, such as *Material Design for Android* or *Human Interface Guidelines for iOS*. We can ask an AI to help us here, but so much of this depends on the specific context and situation that humans need to lead the way.

Edit for Effect

We may also need to edit the AI's output so that it has the right impact. Because we are better than AIs at understanding open-ended situations that require taking context into account, we need to edit the AI's output to make it even more effective for our goal *in context*. For example, if we were trying to nudge a government agency into moving quickly to rescue the stranded ski team, we would refine the message to evoke an emotional response from the audience, such as feeling anxious about the consequences of doing nothing vs. feeling joy and satisfaction in moving quickly to save the day. Or, when writing a speech, we would tailor it to draw in the audience, using rhetorical devices and dynamic language to sway the listeners. Similarly, when creating a short story, we would want to edit it to maximize interest and emotional impact.

In short, AIs can help us deploy creative problem-solving techniques, but we humans need to play a key role in this process.

We close this section with a question that many have asked: Is it "cheating" or "plagiarism" to use the CAL to help us solve problems? What about having an AI help us write something? To address this concern, let me start with an anecdote: At least during the early 2000s, in Paris, France, if you were invited to someone's home for dinner, they typically expected you to bring a bottle of wine or dessert. If you brought dessert, you got "credit" not for having made it yourself, but for knowing where to find a good dessert—for knowing which bakeries made particularly delectable cakes, pies, or the like. Similarly, we can get credit for a product not for having created it entirely by ourselves, but for recognizing when we have induced an AI to produce something of value and then improving it. That is, the ultimate quality of the output of the AI depends on how effectively we, the users, define the goal, write the prompt, evaluate the results, iterate, and ultimately revise and refine the output. At every step, the human user is crucial—the AI wouldn't solve the problem or write the document the way it did without human input. For example, I used the CAL to draft most of the examples in the Tables in this book, but the final product reflects my knowledge, skill, and taste. If employed as described here, the AI really is a cognitive amplifier, and it's the user's cognition that is being amplified.

Thus, provided that we acknowledge that we've engaged in the CAL, the user deserves credit for the final product as a crucial contributor. However, that said, just as it would be dishonest to have said that we made the delicious bakery-bought cake, denying an AI's role in the final product, and implicitly or explicitly claiming to have done *all* the work ourselves, is dishonest. Provided that we are not prohibited from using an AI for a particular purpose, in my view we should get credit for knowing how to use it well.

The idea of cognitive amplifiers has immediate implications for how AIs can help us overcome limitations imposed by the nature of our minds and brains. We next turn to those limitations, and consider how those limitations can impair our using the CAL effectively, and how we can utilize the CAL to help us manage these constraints.

Flourishing with the CAL

We need to become adept at creative problem solving if we are to flourish in the Age of AI. As noted before, learning to work effectively with AIs will be a necessary foundation for having a sense of autonomy and control in the coming years. Moreover, such skills will be required in many and varied types of jobs, and mastering those skills should both contribute to feelings of job satisfaction and help us be valued contributors at work. In addition, we can apply creative problem solving across many domains, which can facilitate our relationships with others. Indeed, as we consider later in this book, creative problem solving can help us grow as people, and can play an important role in helping us discover purpose and meaning in life.

Digging Deeper

To dig deeper into the subject matter, below are examples of videos that address the material reviewed in the chapter. In addition, the below search terms can help the reader locate new videos on the topic.

Creative Thinking: How to Increase the Dots to Connect: https://www.you tube.com/watch?v=cYhgIITy4yY

- "Creative problem-solving techniques"
- "Directed creative thinking strategies"
- "Enhancing creative problem solving with AI"
- "Cognitive processes in creative problem solving"
- "Overcoming cognitive limitations in problem solving"
- "AI-assisted idea generation for problem solving"
- "Evaluating creative solutions with AI"
- "Collaborative problem solving with AI"
- "Context-dependent problem-solving strategies"
- "Refining AI-generated solutions through creative thinking"
- "Divergent and convergent thinking in problem solving"
- "Adapting problem-solving strategies to changing situations"

SCAMPER a creative thinking technique: https://www.youtube.com/watch?v=G8w0rJhztJ4

Convergent Thinking versus Divergent Thinking: https://www.youtube.com/watch?v=cmBf1fBRXms

- "Divergent thinking techniques"
- "SCAMPER method for creative problem solving"
- "Generating alternative ideas through divergent thinking"
- "Enhancing divergent thinking with AI"
- "Substitution technique in creative problem solving"
- "Using analogies for creative idea generation"
- "Reframing obstacles as opportunities in problem solving"
- "Edge cases and creative thinking"
- "Elaborating associations for divergent thinking"
- "Adding constraints to stimulate creative thinking"

What is Convergent Thinking | Explained in 2 min: https://www.youtube.com/watch?v=6a3rF3m9ea4

SWOT Analysis—What is SWOT? Definition, Examples and How to Do a SWOT Analysis: https://www.youtube.com/watch?v=JXXHqM6RzZQ

- "Convergent thinking techniques"
- "Evaluating alternative solutions through convergent thinking"
- "SWOT analysis for decision making"
- "Cost–benefit analysis in problem solving"
- "Identifying and mitigating biases in decision making"
- "Cognitive biases in convergent thinking"

- "Strategies for selecting the best solution"
- "Emotional factors in decision making"
- "Integrating critical thinking in convergent thinking"
- "AI-assisted convergent thinking"
- "Combining divergent and convergent thinking in problem solving"

References

1   Gutman, M. (2018). *The boys in the cave: Deep inside the impossible rescue in Thailand*. William Morrow.
2   Cass S. (2005, April 1). Apollo 13, we have a solution. *IEEE Spectrum Online, 4*, 1. https://spectrum.ieee.org/tech-history/space-age/apollo-13-we-have-a-solution
3   Sarathy, V. (2018). Real world problem-solving. *Frontiers in Human Neuroscience, 12*, 261. doi:10.3389/fnhum.2018.00261.
4   Aboze, B. J. (2024, March 25). How to measure LLM performance. *deepchecks*. https://deepchecks.com/how-to-measure-llm-performance/
5   Wei, F., Chen, X., & Luo, L. (2024). *Rethinking generative large language model evaluation for semantic comprehension*. https://ar5iv.labs.arxiv.org/html/2403.07872v1
6   Zaphir, L., & Lodge, J. M. (2023, June 27). Is critical thinking the answer to generative AI? *Times Higher Education*. www.timeshighereducation.com/campus/critical-thinking-answer-generative-ai
7   Baer, J. (2016). *Creativity and divergent thinking: A task-specific approach*. Psychology Press.
8   Bingölbali, E. & Bingölbali, F. (2020). Divergent thinking and convergent thinking: Are they promoted in mathematics textbooks? *International Journal of Contemporary Educational Research, 7*, 240–252. https://doi.org/10.33200/ijcer.689555
9   Nielsen, D. (2017). *The divergent and convergent thinking book: Notebook for creative thinking*. Lawrence King Publishing.
10 Dam, R. F., & Siang, T. Y. (2024, January 25). SCAMPER: How to use the best ideation methods. *Interaction Design Foundation*. www.interaction-design.org/literature/article/learn-how-to-use-the-best-ideation-methods-scamper
11 Elmansy, R. (2015, April 10). A guide to the SCAMPER technique for design thinking. *Designorate*. www.designorate.com/a-guide-to-the-scamper-technique-for-creative-thinking/
12 Serrat, O. (2017). *Knowledge solutions: Tools, methods, and approaches to drive organizational performance*. Springer. https://doi.org/10.1007/978-981-10-0983-9_33
13 Gentner, D., & Hoyos, C. (2017). Analogy and abstraction. *TopiCS: Topics in Cognitive Science, 9*, 672–693.
14 Holyoak, K. J. (2005). Analogy. In K. J. Holyoak & R. G. Morrison (Eds.), *The Cambridge handbook of thinking and reasoning*. Cambridge University Press, pp. 117–142.

15   Schooley, S. (2024, January 3). What is a SWOT analysis? (And when to use it). *Business News Daily*. www.businessnewsdaily.com/4245-swot-analysis.html

16   Teoli, D., Sanvictores, T., & An, J. (2024). SWOT analysis. In: *StatPearls*. StatPearls Publishing. www.ncbi.nlm.nih.gov/books/NBK537302/

17   Kosslyn, S. M. (2023). *Active learning with AI: A practical guide*. Alinea Learning.

18   Landau, P. (2023, June 21). Cost–benefit analysis: A quick guide with examples and templates. *ProjectManager*. www.projectmanager.com/blog/cost-benefit-analysis-for-projects-a-step-by-step-guide

19   Stobierski, T. (2019, September 5). How to do a cost–benefit analysis & why it's important. *Harvard Business School Online*. https://online.hbs.edu/blog/post/cost-benefit-analysis

20   Acciarini, C., Brunetta, F., & Boccardelli, P. (2021). Cognitive biases and decision-making strategies in times of change: A systematic literature review. *Management Decision*, *59*, 638–652. https://doi.org/10.1108/MD-07-2019-1006

21   Azzopardi, L. (2021). *Cognitive biases in search: A review and reflection of cognitive biases in information retrieval*. Association for Computing Machinery.

22   Caputo, A. (2013). A literature review of cognitive biases in negotiation processes. *International Journal of Conflict Management*, *24*, 374–398. https:doi.org/10.1108/IJCMA-08-2012-0064

23   Ehrlinger, J., Readinger, W. O., & Kim, B. (2016). Decision-making and cognitive biases. *Encyclopedia of Mental Health*. doi:10.1016/B978-0-12-397045-9.00206-8.

24   Rizzolatti, G., & Craighero, L. (2004). The mirror-neuron system. *Annual Review of Neuroscience*, *27*, 169–192.

25   Wellman, H. M., Cross, D., & Watson, J. (2001). Meta-analysis of theory-of-mind development: The truth about false belief. *Child Development*, *72*, 655–684.

26   Saxe, R., & Kanwisher, N. (2003). People thinking about thinking people: The role of the temporo-parietal junction in "theory of mind." *NeuroImage*, *19*, 1835–1842.

Chapter 5

Managing Cognitive and Emotional Constraints

The previous chapters have focused on how to use an AI as a cognitive amplifier by engaging the Cognitive Amplifier Loop (CAL). To deploy the CAL effectively we need more than critical thinking and creative problem solving. We also need to manage cognitive and emotional limitations that interfere with our effective use of the CAL.

In this chapter we focus on key limitations on our thinking that result from the way our brains work. We identify numerous and varied ways to employ the CAL to help us manage those limitations, having it do things that would tax our own capacities.

System 1 and System 2

One way to think about managing our cognitive limitations stems from Daniel Kahneman's highly influential conception of how the mind works.[1] He made a compelling case that the mind relies on two systems (sometimes referred to as "types of processing"; e.g.,[2]): "System 1" carries out evolutionarily older types of mental processing; it is unconscious and automatic, is very fast, and many processes take place simultaneously. For example, if we want to get up to leave the room, System 1 guides us to head for the door and to avoid bumping into things. Similarly, when we are driving and see a red light, our foot automatically lifts off the accelerator and pushes the brake pedal—we don't need to think about this, but rather do it without effort.

In contrast, "System 2" carries out the evolutionarily more recent types of mental processing; it involves conscious evaluation or manipulation, is effortful, slow, and engages in step-by-step thinking. For example, if we bring home several bags of groceries and are struggling to fit the items into the refrigerator, we might visualize moving things that are already in there (e.g., a gallon of milk, head of lettuce, and bag of oranges) in various ways to figure out how to make room for the new arrivals. Or if a

DOI: 10.4324/9781032686653-6

friend tells us something that seems fishy, we might marshal our critical thinking skills to figure out what seems "off." System 2 mental processing often relies on logic, analogy, and other sorts of step-by-step conscious reasoning.

In general, System 2 comes into play when automatic routines don't allow us to cope with a situation. For example, if we drive the same route repeatedly, perhaps to work or to visit a good friend or relative, we eventually can navigate that route by relying on System 1. In fact, we may have no memory at all of the journey when we arrive—System 1 operated like falling dominoes, one turn at a time, completely automatically and unconsciously. Now consider what would happen if a basketball suddenly bounced in front of the car, accompanied by a child's shouts. Now our usual routine would be interrupted, and System 2 would kick in. System 2 would allow us to reason about what to do next—should we just slam on the brakes, steer to the side of the road, pound on the horn, or do something else? A particularly important function of System 1 is to register when automatic routines aren't sufficient and to call on System 2 to take over when necessary.

A central, and crucial, difference between System 1 and System 2 processing is that System 2 relies on *working memory* whereas System 1 does not. Working memory is the "mental workspace" where we consciously interpret and manipulate information.[3,4] When we consciously reason step-by-step, we are relying on working memory. This is important because we can only hold about four distinct pieces of information in working memory at the same time.[5] For example, we can hold about four separate digits, letters, words, or phrases. Notice that what is a "distinct piece of information" is key: If we group letters into a word, the word becomes a whole, a separate entity that we can then store as a single unit. The same applies to grouping words into phrases. But in all cases, whatever the nature of the unit we organize, we can only keep about four of them at our "mental fingertips" in working memory at the same time.

The limited capacity of working memory explains why we should not try to engage in a conversation with someone who is just learning to drive. In the beginning, the novice driver is relying on System 2. With enough practice, what starts off as laborious System 2 work is moved into System 1, where it no longer relies on working memory. But at the outset, the novice driver is stuck with step-by-step conscious thinking in System 2. Because much of their working memory is filled with keeping track of what they are doing while driving, there is less "space" left over for processing a conversation. And if they do decide to process the conversation, they will lose some of the working memory capacity they need to drive the car. Moving information processing to System 1 relieves a burden on working memory because System 1 does not draw on it (cf. [6,7]).

The limitations of working memory affect our ability to use the CAL at every step. However, we can employ the CAL itself to help us cope with this issue. In the following section, we consider a way we can do this.

Using the CAL to Reduce Working Memory Load

The CAL can help us cope with working memory load in System 2 thinking. To illustrate this, I first asked GPT-4 to produce a list of everyday tasks that require System 2 (having explained to it what System 2 does). I then chose one item on the list, namely making financial decisions. Such decision making requires us to take many factors into account, which is clearly taxing if we need to rely on working memory. If we are an expert, we have much of this information in System 1—we recognize familiar scenarios, and immediately know how to respond. But if we are not an expert, such reasoning taxes our System 2 limitations.

One approach to managing working memory limitations in financial decision making is through training. We could recruit an AI to train ourselves on the principles until they become second nature, being shifted into System 1. This would require a lot of training—probably months, or even years—and a lot of work. Yes, an AI can make such training efficient and effective by providing us with individualized feedback, which would allow each of us to zero in on the material we find most difficult. This dynamic, personalized feedback is a huge advantage of using AIs in this way.[8] However, this would still be a lot of work.

Another approach is to learn when to submit a prompt that would have an AI do the heavy lifting, taking advantage of its role as a cognitive amplifier. Here's the key: We should strive to put into System 1 intuitions about when to deploy a specific AI prompt to help us in a particular situation.

The key to moving any process into System 1 is practice. The more we practice, the more the activity becomes automatic, and falls into the purview of System 1. In this example, we would want to consider a wide variety of financial scenarios. If we do this often enough, and the criteria for what counts as a "financial scenario" are crisp and general enough, we will move the decision to engage an AI for this task into System 1.

I asked GPT-4 for the principles people should employ when they evaluate financial situations. I could have simply relied on the training data that went into the AI, but that made me uncomfortable. I wanted to know what the AI would actually be doing, given that I would rely on it to make important suggestions. Using the CAL, I asked the AI to list principles that we should rely on when making financial decisions. It offered a set of principles, which I refined after doing some quick Google searches and using critical thinking. I then incorporated those principles into the following prompt:

Please begin by asking me to give you a financial scenario, which I will want you to help me make a specific judgment about. After I give you the scenario, ask me questions one at a time that you need to have answered in order to use the following principles to make a judgment. Ask me one question at a time, and wait until I respond before asking me another question. Do not continue until I have answered each of your questions, one at a time. After asking me such questions, make a judgment. After you have made your judgment, please explain how you used the principles. Here are the principles to use to make your decision:

1. Risk vs. Reward Assessment

Consider the balance between risk and reward.

2. Prioritization of Financial Goals

Ask me what financial goals are most important to me, which will help you to make decisions that align with those priorities.

3. Understanding of Financial Basics

Rely on basic financial concepts, such as the importance of saving, the power of compound interest, and the basics of diversification.

4. Reaction to Market Trends

Ask about relevant market movements and economic indicators.

5. Use of Heuristics

Rely on simple financial heuristics, or rules of thumb, such as "spend less than you earn," "invest early and often," and "keep an emergency fund."

6. Assessment of Personal Risk Tolerance

Ask me about my risk tolerance, enabling you to take this into account.

7. Spending Discipline

Question the necessity of a purchase, which can help in avoiding impulsive spending and keeping financial goals on track.

8. Emergency Fund Importance

Nudge me to consider maintaining an emergency fund.

9. Debt Aversion for Non-Essential Expenditures

Nudge me to avoid taking on high-interest debt for non-essential expenses.

Here's what happened when I submitted this prompt:

AI:

Please provide a financial scenario that you'd like help making a specific judgment about.

Me:

I'm interested in renting an apartment in London for a week, but the ones in preferred locations are much more expensive than those on the periphery. Should I spend more money on an apartment and less on food and entertainment while I'm there, or should I spend less on an apartment so I can get more out of the city?

AI:

To better assess your situation and help you make a decision, could you please tell me what your primary financial goals are for this trip?

Are they more aligned with comfort and convenience, or are you aiming to maximize your experiences in the city within a budget?

Me:

Comfort and convenience.

AI:

Given your preference for comfort and convenience, how important is it for you to stay in a central location in London? Does being close to major attractions or certain amenities significantly enhance your travel experience?

Me:

It is important, otherwise I'll spend a lot of time just in transit.

AI:

Considering the importance of location to avoid long transit times, how do you feel about potentially spending less on food and enter-tainment? Would you be comfortable with more budget-friendly options in these areas to accommodate a better apartment location?

Me:

I could live with that, but I would prefer to take advantage of good ethnic restaurants in London.

AI:

Given your interest in exploring ethnic restaurants, which could be a significant part of your travel experience, how would you describe your overall risk tolerance when it comes to managing your budget on this trip? Are you comfortable with potentially stretching your budget for a central apartment and still trying to enjoy quality dining, or would you prefer a more cautious financial approach?

Me:

I have to work within a strict budget.

AI:

Given your strict budget constraints and the importance of both location and dining experiences, it might be best to find a balance. Here's my suggestion based on the principles provided:

1. Risk vs. Reward Assessment: Choosing a central location mini-mizes the risk of wasting time and potential transportation costs, which aligns with your preference for comfort and convenience.
2. Prioritization of Financial Goals: You've indicated a strong prefer-ence for comfort and enjoying ethnic restaurants. Thus, spending more on a centrally located apartment could be justified as it aligns with your goals, allowing easier access to preferred dining.
3. Understanding of Financial Basics: It's crucial to stay within your budget to avoid financial stress. This suggests looking for deals or less expensive but well-located accommodations.
4. Reaction to Market Trends: Keep an eye on rental prices as they can fluctuate. Booking early might help you secure a better rate.
5. Use of Heuristics: Spending less on some meals or choosing budget-friendly dining options occasionally can allow you to enjoy a few higher-end meals without exceeding your budget.
6. Assessment of Personal Risk Tolerance: Staying within a strict budget means managing expenses tightly. Opting for an apart-ment that is slightly less expensive but still central could free up funds for dining.
7. Spending Discipline: Prioritize your spending on what matters most—accessibility and food, and maybe cut back on other less essential expenditures during the trip.
8. Emergency Fund Importance: Ensure you have a small buffer in your budget for unexpected expenses.
9. Debt Aversion for Non-Essential Expenditures: Avoid overspend-ing on luxury accommodations or high-end meals that could push your budget beyond its limits.

> In conclusion, I would recommend finding a moderately priced apartment in a central location. This way, you can still enjoy the city's ethnic cuisine without compromising too much on comfort or overspending. This approach respects your financial goals and constraints, ensuring a balanced and enjoyable experience.
>
> **Me:**
>
> Thank you.
>
> **AI:**
>
> You're welcome! If you have any more questions or need further assistance as you plan your trip, feel free to ask. Have a great time in London!

An AI is a cognitive amplifier, which can help us reason systematically about a topic we care about. If we rely on the CAL in this way often enough, we will automatically want to consult an AI when faced with such situations. By so doing, we have the AI do much of the heavy lifting, but under our control. Indeed, if we employ this process often enough, we will start picking up the principles that the AI is using, and eventually may shift them into System 1—and thus not need to use the AI as often.

Managing Emotion

As if the built-in limitations of working memory weren't bad enough, our emotions can make them even more severe. "Emotions" are subjective responses that (a) have a "valence," ranging from positive (such as joy) to negative (such as despair); (b) have a strength (joy and euphoria are both positive, but the latter is stronger); (c) are usually associated with bodily states (such as increased heart rate and sweaty palms); and (d) typically produce changes in behavior (leading us to approach or avoid the focus of the emotion). I asked GPT-4 the following question: "Can you feel emotion?" It responded: "I don't have emotions or consciousness. I process and generate text based on patterns and information in the data I was trained on. My responses are generated based on a combination of learned information and the specific instructions given to me in the query, without any personal experience or emotional understanding." If we ask a human the same question, they probably will look puzzled and wonder why we are asking such a silly question. Of course humans feel emotions!

Our emotions affect our cognition and behavior in many ways. Some of these ways can help us and some can hurt us, but all of them affect how we interact with each other—and will no doubt affect how we use the CAL when we interact with AIs.

Negative Effects of Emotions on Cognition

Emotions can impair many aspects of cognition, especially those that rely on System 2 processing. In what follows we briefly consider key cognitive functions that may be undermined by emotion. We then consider ways to manage our emotions, so that we can reduce these unfortunate effects.

Attention. Some of the detrimental effects of emotion are a consequence of how it affects our attention. When we feel strong negative emotions, our attention narrows and we focus on a single thing to the exclusion of almost everything else.[9] For instance, if a mugger were to pull a gun on you, you would probably focus on the gun, and not notice anything else—including the mugger's face. This would make it difficult to later recall what they looked like. In general, emotional states alter which objects or events we tend to notice. For instance, if we feel sad we are more likely to notice negative objects or events, but if we feel happy we are more likely to notice positive objects or events.[10]

Memory. Emotion can also have a profound impact on memory. Emotional events are usually remembered better than non-emotional ones. We are especially likely to remember events that evoke negative emotions, probably in part because they draw our attention to them at the time the events unfolded. In addition to that, however, our brains appear to prioritize storing such information.[11] Some of this effect may be produced by the arousal associated with emotional experiences: When we are more aroused, this enhances the vividness of memory. We tend to remember events that are emotionally intense.[12]

Reasoning and decision making. Any strong emotion can disrupt our reasoning and decision making. For example, if you found out you won the lottery, do you think you could reason clearly about how to spend the funds as soon as you got the good news? Researchers have found that negative emotional content in particular can interfere with our ability to evaluate logical arguments.[13] However, the ways we falter may depend on the particular emotion. For example, fear may make us less likely to take risks whereas anger may make us more likely to do so.[14]

In addition, emotions can undermine reasoning and decision making by fueling many of the biases summarized in Table 4.3. In fact, emotion can intensify the search for information or interpretations that support what we already believe or want to believe, thereby increasing confirmation bias.[15] Similarly, emotional investment can enhance the sunk cost fallacy, where we continue a course of action because of past investment in it. We are also prone to an "Emotional Framing Effect," where our decisions are influenced by how information is emotionally expressed, rather than just by the content.[16] We typically prefer options that are framed positively.

Such effects of emotion can distort convergent thinking during creative problem solving. In addition, we can develop a strong emotional attachment to a particular candidate solution, regardless of its likely efficacy. This can lead us to select a less optimal solution because we became emotionally attached to it.

Threat response. The *fight, flight, or freeze response* is another example of how emotion can affect cognition and behavior (e.g.,[17,18]). This response is a bodily reaction to a perceived threat, which consists of a slew of hormonal changes and physiological reactions. The response prepares the body either for rapid action, when we confront or flee from a perceived threat, or for freezing in place. For instance, if we are confronted by a mugger and want to fight or flee, we will quickly experience a raft of physiological effects. Most notably, we generally will be more aroused, which decreases our reaction time; our heart rate will increase, which pumps more blood to vital organs and muscles; we will breathe more rapidly and shallowly, which increases oxygen intake; our pupils will dilate, which allows more light into the eyes, thereby improving vision. In contrast, if we experience the freeze response to threat, our heart rate drops, our muscles tense but we don't move. We become very still and attentive to what is going on around us. These fight, flight, or freeze responses apparently were a critical survival mechanism for our ancestors, which allowed them to react quickly to life-threatening situations.

In today's world, stressors that are not life-threatening, such as work pressure or financial worries, can trigger the fight, flight, or freeze response. This unnecessary response can lead to chronic stress—which in turn can produce health issues.[19,20] Furthermore, the stress response can also lead to rumination and worrying, which disrupts working memory and thereby undermines System 2 processing.[21,22] High-stress situations can lead to quick, emotion-driven decisions rather than careful, rational deliberation. Stress can result in impulsive choices or a narrow focus on short-term gains over long-term benefits.

In short, it behooves us to manage our emotions in order to flourish in the Age of AI. Research shows that we are not slaves to our emotions and we can mitigate them in various ways. We next consider ways to deploy an AI to help us manage our emotions.

Managing Negative Emotions Through Meditation

One way to cope with negative emotions is to rely on Mindfulness-Based Stress Reduction meditation (MBSR[23]). MBSR is an evidence-based training program that relies on mindfulness meditation to help people manage stress, anxiety, depression, and pain—and, in so doing, also improve their ability to manage working memory. It's typically an eight-week program that includes weekly group sessions and daily mindfulness practices at home. Consider these components of MBSR:

1. *Body Scan Meditation.* Participants lie down and focus their attention on various parts of their body a part at a time, starting from their toes and moving upwards.
2. *Sitting Meditation.* Participants sit quietly and focus on their breath, a mantra, or sensations in their body. When their attention wanders, they notice where it has gone and gently redirect it back to the focus of the meditation.
3. *Walking Meditation.* Walking with attention to the sensations of the feet touching the ground can help improve focus and awareness.

Participating in this MBSR program does in fact enhance attention and working memory. Indeed, participants in one study improved their attention spans after only four days of MBSR.[24] Such brief training also improved working memory capacity, which may have been a byproduct of reducing the number of distracting thoughts. Indeed, such meditation training can help people sustain attention for a longer period of time.[25] One result of this increased capacity is that people can process information more efficiently, which in turn allows them to switch to the next task more quickly. Consistent with these behavioral findings, after only eight weeks in a MBSR program, MRI scans revealed increased gray matter in brain regions involved in learning and memory, emotion regulation, self-referential processing, and perspective-taking.[26]

We can employ an AI to support and enhance the practice of MBSR. Unlike existing apps, an AI can respond to our needs and wishes dynamically, and thus tailor the way it delivers the procedure differently, depending on our situation and state of mind. For instance, we can ask it to fit the mediation into the amount of time we have available, we can ask it to speed up or slow down, and we can ask it to focus on some specific aspects of the procedure and not others. Employing an AI allows us to make such adjustments quickly and easily, changing them to fit our current situation and mood.

For example, I had GPT-4 help me create the following prompt, which we can use to guide meditation—ideally with the voice options turned on. That is, set your LLM so that it reads the text aloud, so you can close your eyes and listen. And set it to receive voice commands, so you don't need to type into your device. Select a voice that you find soothing, which may not be the same voice that others prefer. Load the following prompt into the AI, and respond "Ready," "Yes," "Go," or the like whenever it asks you to let it know that you are ready to proceed to the next step. "You" in this case refers to the LLM:

You are going to help me conduct Body Scan Meditation. Begin by saying everything between quotation marks and following the requests that are not within quotation marks. Please begin by saying:

"Welcome to your personalized Body Scan Meditation session. Before we begin, please let me know how much time you have available for this exercise. This will help me tailor the session to fit your schedule perfectly.

How much time can you dedicate to today's Body Scan Meditation?

• Less than 10 minutes.
• 10–20 minutes.
• More than 20 minutes."

Please pause at this point and wait for me to respond. Do not provide an answer for me, and wait until I give you one of the three choices above. Once I have responded, continue by telling me the following:

"During the session, you will be guided to focus on different parts of your body. We'll start at your toes and gradually move upwards towards the crown of your head. Please tell me when you are ready to move on to the next step.

If at any point you feel like the pace is too fast or too slow, or if you wish to spend more time focusing on a particular area of the body, please tell me:

• Speed up: If you'd like to go through the body parts more quickly.
• Slow down: If you'd like to spend more time on each body part.
• Repeat [body part]: If you'd like to revisit any part of the body.

Now, find a comfortable place to lie down, and let's begin when you're ready."

If I ask you to speed up, please go through the parts more quickly. If I ask you to slow down, please slow down your pace of going through the body parts. If I ask you to repeat, please revisit the last body part.

Now resume by saying the following. If at any time the user asks you to speed up, slow down, or repeat, please do so.

"Setting the Scene:
First, find a quiet and comfortable place where you can lie down without interruptions. You may use a yoga mat, a carpeted floor, or any flat surface that supports your body comfortably. Use a pillow under your head or knees if needed. Once you're lying down, take a few deep breaths, in through your nose and out through your mouth, allowing your body to start relaxing.

Starting at the Toes:
Now, bring your attention to your toes. Notice any sensations you feel here—perhaps warmth, coolness, tingling, or maybe no sensation at all. Take a deep breath in, and as you breathe out, imagine any tension in your toes being released and flowing out of your body. Tell me when you are ready to go to the next step."

At this point pause. Do not proceed until the user tells you that they are ready to go to the next step. At that point, please read the below.

"Moving Up to the Feet and Ankles:
As you continue to breathe deeply and evenly, shift your focus to your feet and ankles. Pay attention to how they feel against the surface beneath you. With each exhale, let go of tension and relax a little more."

At this point pause. Do not proceed until the user tells you that they are ready to go to the next step. At that point, please read the below.

"Progressing to the Lower Legs:
When you're ready, move your awareness up to your calves and shins. Notice the muscles in this area. Are they tight or relaxed? As you breathe in, imagine drawing in fresh energy, and as you breathe out, let the muscles soften and release any tightness."

At this point pause. Do not proceed until the user tells you that they are ready to go to the next step. At that point, please read the below.

"Acknowledging the Knees and Thighs:
Gently bring your attention to your knees and then up to your thighs. This area supports much of your daily activities, so it might hold tension. Breathe into your thighs, and on each exhale, feel them sinking deeper into relaxation."

At this point pause. Do not proceed until the user tells you that they are ready to go to the next step. At that point, please read the below.

"Focusing on the Hips and Pelvis:
Now, move your awareness to your hips and pelvis. This is a central area that can often hold stress. Inhale deeply, filling this area with a sense of calm, and as you exhale, imagine the stress melting away."

At this point pause. Do not proceed until the user tells you that they are ready to go to the next step. At that point, please read the below.

"Climbing to the Torso:
Direct your attention to your lower back and abdomen. Feel the rise and fall with each breath. Next, focus on your upper back and chest. If you notice any areas of tightness, use your breath to soften them."

At this point pause. Do not proceed until the user tells you that they are ready to go to the next step. At that point, please read the below.

"Visiting the Hands and Arms:
Shift your focus down to your fingers, hands, and wrists. Then, move up through your forearms, elbows, upper arms, and shoulders. With each breath, release any tension you carry in these areas."

At this point pause. Do not proceed until the user tells you that they are ready to go to the next step. At that point, please read the below.

"Reaching the Neck and Head:
Finally, bring your attention to your neck, which often holds a lot of tension. Take a moment to breathe into your neck, letting the muscles relax. Then, scan your face—your jaw, cheeks, eyes, forehead, and the top of your head. Allow your whole face to soften."

At this point pause. Do not proceed until the user tells you that they are ready to go to the next step. At that point, please read the below.

"Concluding the Meditation:
To finish, take a few more deep breaths, feeling the weight of your relaxed body against the surface beneath you. When you're ready, gently wiggle your fingers and toes, slowly awakening your body. Open your eyes when it feels right, and slowly sit up, taking a moment to notice the calmness and relaxation you've cultivated."

Mindfulness practices derived from MBSR can influence our cognitive processes by shifting some tasks from an effortful, step-by-step System 2 process to a more automatic, unconscious System 1 process. We next consider how such a transition might work.

Reframing Cognitions to Reshape Emotions

Many methods of managing disruptive emotions rely on "reframing." We can deliberately change our interpretation of a situation in order to modify our emotional response in positive ways. For example, we might cope with anger in a traffic jam by reinterpreting the situation as an opportunity to listen to a favorite podcast, which would thereby reduce our anger.

"Cognitive restructuring" is a tried-and-true method to help people reappraise the causes of troubling emotions (e.g.,[27,28]). This method allows people to reinterpret a situation so that they can replace negative or irrational thoughts with more balanced and realistic ones—which reduces the emotional distress caused by the initial thoughts. Key steps of this process include the following:

- *Identifying the problem.* The first step is to recognize the thoughts that produce emotional distress.
- *Examining and challenging the evidence.* We next need to question whether these thoughts are accurate and useful.
- *Replacing inaccurate thoughts.* Finally, we need to replace the negative or irrational thoughts with more accurate or positive ones.

Research indicates that this approach does in fact help people recast their emotions and behaviors, reducing the effects of negative emotions (e.g.,[29]). We can utilize an AI to teach this method, but this requires a lot of work. However, in some situations, we can have the AI do some of the work for us, which can potentially help us to cognitively restructure a disturbing situation. Indeed, researchers report that GPT-4 often does a better job of reappraising negative situations than do humans.[30]

Try the following prompt:

Please ask the user his or her name. Pause and wait for the user to respond. Do not proceed until the user has entered their name. After this, tell the user: "We are going to work together to consider whether certain negative thoughts and interpretations of situations are accurate. And if they are not, then the resulting negative emotion is not appropriate. To deal with such emotion, we will work together to think of better ways to regard that situation."

Following this, ask the user whether they are ready to go on. Wait for their response. If the response is positive, continue to Part 1 below. If it is negative, tell the user: "OK. Please feel free to return at another time and we can continue."

Part 1: Identifying the Origins of Negative Emotions
At the outset of Part 1, tell the user: "Think about a recent situation where you felt upset, anxious, or distressed. What were the thoughts running through your mind at that moment? Write them down as accurately as you can. Try to pinpoint the exact thoughts that made you feel the worst."

Wait until the user responds. Do not proceed until they have entered their response. Following this, proceed to Part 2.

Part 2: Examining and Challenging the Origins of Negative Emotions
At the outset of Part 2, you are going to ask the user three questions. Pause after each question and allow the user to respond. Do not present all three questions at once; present them one at a time, and only continue to the next one after the user responds. Begin by telling the user the following: "Now, take each thought or interpretation you've identified and ask yourself the following three questions. Write down your answers. This process helps you to see other perspectives and assess the validity of your initial thoughts. Here are the three questions:

1. What concrete evidence supports this thought or interpretation? Is this solid?
2. In what other ways could you interpret the situation? Are there other interpretations that would not lead to the negative emotion?

3. In what ways can you imagine that someone else might view this situation differently?"

Part 3: Replacing the Origins of Negative Emotions
At the outset of Part 3, you are going to tell the user the following: "Based on your examination and challenge of each thought or interpretation, now try to formulate more balanced and realistic thoughts. For each negative thought, write a more constructive or positive perspective. Consider what you would say to a friend in a similar situation. Aim for thoughts that are not overly optimistic but grounded in reality. How does acknowledging these more balanced thoughts make you feel?"

Wait for the user to respond. Following this, ask "Is there anything else you would like to do to help reinterpret the thoughts and interpretations that have led to negative emotions? Would you like to return to them and rethink any of the above? If so, we can start again—just let me know. If not, I wish you well and am happy to help in the future."

If the user replies that they would like to revisit the above material, start from the beginning of Part 1 and repeat the process. Otherwise, close the session.

Here's what happened when I tried it out, focusing on a low-emotion, rather innocuous personal issue:

AI:

Could you please tell me your name?

Me:

Stephen

AI:

Thank you, Stephen. I am going to work with you to see that certain negative thoughts and perceptions can be reinterpreted. These reinterpretations will allow you to see the situation differently, and to feel better about it.

Are you ready to go on?

Me:

Yes

AI:

Great, let's get started.

Please tell me about the situation that has led you to feel upset, anxious, or distressed. What thoughts are running through your mind? Please try to pinpoint the exact thoughts that make you feel the worst, and write them down as accurately as you can.

Me:

Someone whose opinion I value has stopped responding to my emails, and I cannot help that I insulted him in some way.

AI:

Thank you for sharing that, Stephen. It sounds like a challenging situation.

Now, let's take the thought that you might have insulted this person. Ask yourself the following question: What concrete evidence supports this thought or interpretation? Write down your answer, please.

Me:

The only evidence I have is that the last time I saw him I had to leave early and he seemed miffed. And then, of course, he's stopped responding.

AI:

Thank you for providing that context. It's understandable how those events could lead you to feel that way.

Based on your examination of the evidence, I'll now formulate a more balanced and realistic version of this thought. Instead of thinking that you definitely insulted him, a more neutral interpretation could be: "He might have been a bit disappointed when I had to leave early, but there could be many reasons why he hasn't responded yet. It's possible he's been busy or dealing with his own issues."

How does acknowledging this more balanced version make you feel?

Me:

I'm skeptical. I've sent a couple of emails and gotten no response. I think I offended him.

AI:

It's natural to feel skeptical, especially when you're worried about the status of an important relationship. Let's consider another angle: if you offended him, it's also possible that he may need some time to process or he might be caught up in other aspects of his life, affecting his response time.

How about this interpretation: "Even if my departure was abrupt, it's possible he's dealing with other priorities or challenges right now. My emails might have been overlooked or are lower on his list of urgencies."

Does thinking about it this way help ease some of your concerns, or do you feel there might be a different approach that could help you view this situation in a less distressing light?

Me:

I think that's right that he has deprioritized my email. But that's indicative of the problem. I wish I had been more considerate at the time and explained in more detail why I had to rush off.

AI:

Reflecting on how you could have handled the situation differently shows a lot of self-awareness and a desire to maintain good relationships, which is commendable.

Here's a reinterpretation that might help: "Although I had to leave abruptly last time, it's an opportunity for me to communicate more openly in the future. My intention wasn't to offend, and reaching out to explain why I had to leave might help clear up any misunderstanding."

This approach not only acknowledges your feelings but also focuses on positive steps you can take to possibly mend the situation. How does this perspective make you feel about the situation now?

Me:

I like the proactive approach. Thank you.

AI:

You're welcome, Stephen! I'm glad you found the proactive approach helpful. It's great to focus on constructive steps you can take to improve the situation and communicate effectively.

Is there anything else you would like to do to help reinterpret the thoughts and interpretations that have led to negative emotions? Would you like to return to them and rethink any of the above? If so, we can start again—just let me know. If not, I wish you well and am happy to help in the future.

Notice that the AI did not always incorporate verbatim the specific wording provided in the prompt, but rather sometimes distilled the underlying intention and expressed it differently. If we had wanted the AI always to use the precise wording in the prompt, we could revise the prompt to emphasize that requirement and tell it explicitly not to paraphrase or improvise. If you try out this prompt, you may be surprised by how constructive and empathic the AI seems. Indeed, generative AI appears to have great promise for providing psychotherapy (e.g.,[31,32,33]). The AI is not judgmental and seems to be receptive to whatever we care to say. However, keep in mind that the AI is just anticipating which words follow from which words—it doesn't really "understand" what we write. But, that said, as a cognitive amplifier it can help us manage our cognitive and emotional limitations, in this case our blind spots and inability to take a step back and consider our situation from a more objective perspective. As usual, however, we should view the AI's suggestions with more than one grain of salt.

Managing Moods

A "mood" is a temporary, low-intensity emotional state—such as feeling gloomy or tranquil. Even such relatively weak emotions can affect our judgments. For example, when we are in a positive mood we are more likely to make optimistic judgments and evaluations, whereas when we are in a negative mood we are more likely to make pessimistic judgments and evaluations.[34] And mood affects more than just our judgments. Gordon Bower and colleagues demonstrated that our memories are affected both by the mood we are in when we first encounter an object or event and the mood we are in when we later recall it. We are more likely to recall information that is congruent with our current mood. For instance, if we are happy, we are more likely to remember positive events, whereas if we are sad, we are more likely to remember negative events.[35]

Moods affect us every day, sometimes motivating us to charge ahead and sometimes dragging us down. Given this fact, it's useful to have some control over our moods. *Mood Induction Procedures* (MIPs) are methods that evoke specific moods. Some of these methods are mundane, such as listening to specific types of music, visualizing certain situations, watching particular films, or reading specific emotional descriptions. Other methods are more esoteric, such as hypnotizing people to lead them to feel specific moods. Researchers have validated these procedures by showing that the induced moods affect what sorts of memories come to mind, in accordance with the congruence effects noted above.

Let's consider how we can use mental images to affect our moods (e.g.,[36,37,38]). I asked GPT-4 to provide an imagery-based script that would be an MIP for a calm and relaxing mood, and it produced the following. We would ask the AI to utilize text-to-speech routines to read this aloud, and ask it to speed up or slow down as appropriate. After so instructing it, we would close our eyes, listen, and form the mental images of the scenes it describes.

Setting: A serene beach at sunset.

Begin:
Imagine yourself walking along a secluded beach. The sun is setting, painting the sky in shades of orange, pink, and purple. The air is warm with a gentle breeze that caresses your skin. You can hear the rhythmic sound of waves gently crashing against the shore, a soothing, repetitive sound that calms your mind.

With every step you take on the soft, white sand, you feel more relaxed. The sand is warm under your feet, providing a comforting contrast to the cool breeze. You can smell the salt in the air, a fresh and clean scent that fills your lungs with every deep breath you take.

As you walk further along the beach, you find a comfortable spot to sit and watch the sunset. The sand molds to your body, offering a natural seat. You can feel the warmth of the sand against your skin, grounding you in this peaceful moment.

Looking out over the water, you notice the colors of the sunset reflecting on the surface, creating a mirror image of the sky. The sound of the waves becomes more pronounced, a symphony of nature that is both powerful and peaceful. You take a deep breath in, tasting the salty air on your tongue, and as you exhale, you feel any tension in your body melting away.

The breeze carries the faint scent of tropical flowers from somewhere nearby, adding a floral note to the salty air. You close your eyes for a moment, focusing on the sensation of the breeze and the sound of the waves. In this moment, you are completely at peace, enveloped in the beauty of nature.

As the sun dips below the horizon, the sky turns to deeper shades of red and purple, and the first stars begin to appear. The beach gradually darkens, but the sound of the waves remains constant, a reminder of the world's natural rhythms. You feel a deep connection to the earth, a sense of calm that permeates your being.

End:
When you're ready, you slowly stand up, feeling the sand slip through your fingers as you brush it off. You take one last look at the horizon, now twinkling with stars, and carry with you the sense of calm and relaxation back into your daily life.

This script is designed to be read slowly, allowing the listener or reader to fully immerse themselves in the sensory details and experience a deep sense of relaxation.

What if you aren't a "beach person"? Perhaps you are disquieted by the mere thought of being near the wide-open sea. This would be a problem if we were using a pre-recorded tape to guide us, but is not a problem with an AI: All we need to do is ask it to use another context. In fact, we can ask the AI to give us a slate of choices, perhaps laying down in a meadow next to a trickling brook, relaxing in a hammock strung between two trees, or stretched out on a sofa on the porch of a grand Victorian house situated down a country road. The AI can easily adapt the content to fit our preferences. Indeed, if one of these contexts isn't working well, we can ask it on the spur of the moment to switch to another.

In addition to using a MIP, we can have an AI help us engage in mindfulness meditation as a way to influence our moods. Researchers have reported that people who practice mindfulness meditation regularly have reduced anxiety and depression, both of which produce negative moods (e.g.,[39,40]). This type of meditation leads to positive changes in mood, which in turn decreases the number of negative memories that bubble to the surface.

A Window into the Unconscious

Emotions give rise to useful "hunches,"[41] as discussed in Chapter 1, and can facilitate our reasoning and decision making more generally. For example, emotions can help us prioritize what is important when we reason. If an event evokes a strong emotion, we may interrupt what we are doing and refocus on that event. For instance, if we see someone else in distress, we may feel empathetic concern, which motivates us to help.

These roles of emotion grow in large part out of its effects on our unconscious. We can enhance such contributions of emotion to reasoning and decision making by letting our emotional reactions sit, "sleeping on them." Doing so gives the brain a chance to integrate our short-term reactions into a broad context. Indeed, waiting to respond until after a good night's sleep gives the brain a chance to perform such integration (e.g.,[42,43]).

Because these roles of emotion are unconscious, by definition we are not aware of them until they influence us. However, we sometimes want to predict our responses, so we can either prepare or consider changing our mind. In such situations, it behooves us to know in advance how we are going to feel about a particular decision. There's a simple way to tap into the unconscious body of digested information that affects how we reason and make decisions. Let me illustrate with an example, needing to choose between two jobs. We consider the different facets of each, reacting to some positively and some negatively. The trade-offs are vexing, and we twist and turn in our thinking. So we sleep on it. In the morning, we simply flip a coin—heads we take one job, tails the other. The key is to observe our emotional reaction to the outcome of flipping the coin. Are we relieved? Did we have a negative jolt? This reaction is tapping into the unconscious work of the brain, and can help us surface how we actually feel about the situation. The goal of flipping the coin is to notice how we respond when the "choice" is temporarily taken out of our hands, and to use that information to make our actual decision.

Similarly, we can engage in the CAL and notice how we feel about the results. For example, we can utilize this prompt:

> You are playing the role of a wise advisor. I have a dilemma, and am wrestling with the outcome. Please ask me to describe the dilemma. Wait for me to respond before continuing; do not move on until I have responded. After I respond, consider what I say and then make a decision for me. After this, ask me how I feel about that decision—whether I agree or disagree.

Here's what happened when I submitted this prompt:

AI:

Please describe your dilemma.

Me:

I have been asked to attend a meeting in London, but the date is inconvenient. I would probably learn some useful things at the meeting, and meet some interesting people. But I would need to move my schedule around, which would irritate some people. I can't decide whether the benefits of going would outweigh the drawbacks.

AI:

Given the scenario, I recommend attending the meeting in London. The potential to gain new knowledge and expand your network often presents long-term benefits that can outweigh the short-term inconvenience of rescheduling. While it may cause some initial irritation, those affected will likely understand the importance of this opportunity for your professional growth.

How do you feel about this decision—do you agree or disagree?

As it turned out, I was delighted by its response. I apparently wanted to go to London, which was not so obvious to me before I enlisted the help of the AI.

Flourishing with the CAL

In this chapter, we have seen that we can deploy the CAL to address our cognitive and emotional limitations by offloading cognitive burdens to the AI, such as by having it help us reason through a financial decision or help us reframe troubling situations. Using the CAL in such ways clearly can increase our sense of autonomy and control. Similarly, addressing such limitations can also lead us to be more productive at work, which—all else being equal—should increase our job satisfaction. Indeed, by helping us manage our limitations, the CAL enables us to grow as individuals and holds promise of helping us to develop our abilities and talents.

Moreover, coming to grips with our emotions provides insights that facilitate our relationships. The following chapter begins Part II of the book, which focuses on how to use everything that we've considered so far to improve how we interact with other people. To begin, we need to recognize that different people often may behave differently in the same situation. In many cases, such idiosyncratic responses create a unique event that requires creative problem solving, not simply falling back on well-worn responses. One reason that people respond differently is that they have different personalities. We turn to this topic in the next chapter, which then sets us up to see how we can utilize the CAL in a wide range of social contexts, helping us to flourish in a changing world.

Digging Deeper

To dig deeper into the subject matter of this chapter, consult the sorts of videos noted below. In addition, the search terms can help the reader locate new videos on the topic.

Thinking, Fast and Slow | Daniel Kahneman | Talks at Google: https://www.youtube.com/watch?v=CjVQJdIrDJ0
Daniel Kahneman: Thinking Fast and Slow, Deep Learning, and AI | Lex Fridman Podcast #65: https://www.youtube.com/watch?v=UwwBG-MbniY

- "Kahneman's System 1 and System 2 thinking"
- "Working memory limitations and cognitive performance"
- "Moving from System 2 to System 1 thinking"
- "AI-assisted financial decision making"
- "Developing financial intuitions through practice"
- "Cognitive principles for financial decision making"
- "AI-generated scenarios for learning financial principles"
- "Internalizing financial decision-making principles"

- "Automating financial decision making through AI"
- "Gamification of financial decision-making training"

Peter Doolittle: How your "working memory" makes sense of the world: https://www.youtube.com/watch?v=UWKvpFZJwcE
   Chunking: Learning Technique for Better Memory: https://www.youtube.com/watch?v=hydCdGLAh00

- "Working memory capacity and limitations"
- "Strategies for overcoming working memory limitations"
- "Chunking information for better memory"
- "Organizing information into higher-order units"
- "AI-assisted working memory training"
- "Gamification of working memory strategies"
- "Grouping strategies for remembering information"
- "Cognitive strategies for managing working memory"
- "Techniques for improving working memory capacity"
- "Role of attention in working memory"
- "Mnemonic devices for organizing information"
- "Overcoming cognitive limitations with AI"

6 Major Theories of Emotion | Psych Nerd | Psychology: https://www.youtube.com/watch?v=UOTp7RHD0vs
   The science of emotions: Jaak Panksepp at TEDxRainier: https://www.youtube.com/watch?v=65e2qScV_K8

- "Bodily reactions and emotional experience"
- "Interpretation of bodily reactions in emotions"
- "Situational context and emotional appraisal"
- "Multiple neural mechanisms of emotion"
- "Amygdala and emotional processing"
- "Approach vs. avoidance emotions"
- "Social and cultural factors in emotion"
- "Subjective construction of emotional experiences"
- "Cognitive strategies for modulating emotional responses"
- "Neuroscience of emotion regulation"
- "Schachter-Singer theory of emotion"
- "Lazarus cognitive appraisal theory of emotion"

Introduction to Cognition and Emotion: https://www.youtube.com/watch?v=wIgYPPpuRVk
   Reason vs. Emotion: Two Systems at War? https://www.youtube.com/watch?v=qGMg3w_FC3c

- "Emotion and attention"
- "Mood-congruent perception"
- "Emotional arousal and memory enhancement"
- "Emotion and reasoning strategies"
- "Fear, anger, and risk assessment"
- "Emotional intensity and decision making"
- "Emotions as guides in decision making"
- "Emotion prioritization in reasoning"
- "Emotional intuition and decision making"
- "Integrating emotions in decision making"
- "Ventromedial prefrontal cortex and emotional processing"
- "Unconscious emotional processing and decision making"

What are Cognitive Distortions? https://www.youtube.com/watch?v=otFr NM7PnME

What is Cognitive Restructuring? https://www.youtube.com/watch? v=nRl1TQWrSHw

- "Cognitive restructuring techniques"
- "Reframing negative thoughts and emotions"
- "Identifying and challenging cognitive distortions"
- "Examining evidence for negative thoughts"
- "Replacing negative thoughts with balanced perspectives"
- "AI-assisted cognitive restructuring"
- "Chatbot-guided emotional reappraisal"
- "Cognitive-behavioral therapy for emotional regulation"
- "Thought records for cognitive restructuring"
- "Socratic questioning in cognitive restructuring"
- "Cognitive reappraisal strategies"
- "Emotional distress and cognitive restructuring"

Cognitive Distortions: The Mood Congruent Memory Bias: https://www.youtube.com/watch?v=OGc7FdCoy1Y

The Scientific Power of Meditation: https://www.youtube.com/watch? v=Aw71zanwMnY

- "Mood and judgment"
- "Mood congruence effects on memory"
- "Mood Induction Procedures (MIPs)"
- "Mental imagery and mood induction"
- "Guided imagery for relaxation"
- "Mindfulness meditation and mood regulation"
- "Mindfulness and emotional well-being"

- "Mindfulness-based stress reduction"
- "Mood and cognitive performance"
- "Emotion regulation strategies"
- "Mood and decision making"
- "Mood and information processing"

The Scientific Power of Meditation: https://www.youtube.com/watch?v=Aw71 zanwMnY

Using Artificial Intelligence to Improve Mindfulness and Evaluation Practice: https://www.youtube.com/watch?v=aK2kfJZHpy4

- "Human attention limitations"
- "Multitasking and cognitive performance"
- "Brain-training games for attention and working memory"
- "Transfer of learning in brain-training"
- "Mindfulness-Based Stress Reduction (MBSR) benefits"
- "MBSR and attentional improvements"
- "Mindfulness meditation and working memory capacity"
- "Neuroplasticity and mindfulness training"
- "AI-assisted mindfulness practices"
- "AI-guided body scan meditation"
- "Automated mindfulness coaching with AI"

References

1   Kahneman, D. (2011). *Thinking, fast and slow*. Farrar, Straus and Giroux.
2   Evans, J. S. B. T., & Stanovich, K. E. (2013). Dual-process theories of higher cognition: Advancing the debate. *Perspectives on Psychological Science, 8*, 223–241.
3   Baddeley, A. D. (2007). *Working memory, thought and action*. Oxford University Press.
4   Baddeley, A. D. (2010). Working memory. *Current Biology, 20*, R136–R140.
5   Cowan, N. (2001). The magical number 4 in short-term memory: A reconsideration of mental storage capacity. *Behavioral and Brain Sciences, 24*, 87–114.
6   Sweller, J. (1988). Cognitive load during problem solving: Effects on learning. *Cognitive Science, 12*, 257–285.
7   Sweller, J. (1999). *Instructional design in technical areas*. Australian Council for Educational Research.
8   Kosslyn, S. M. (2023). *Active learning with AI: A practical guide*. Alinea Learning.
9   van Steenbergen, H., Band, G. P. H., & Hommel, B. (2011). Threat but not arousal narrows attention: Evidence from pupil dilation and saccade control. *Frontiers in Psychology, 2*, 281. doi:10.3389/fpsyg.2011.00281.

10   Phelps, E. A., Ling, S., & Carrasco, M. (2006). Emotion facilitates perception and potentiates the perceptual benefits of attention. *Psychological Science*, *17*, 292–299.

11   LaBar, K. S., & Cabeza, R. (2006). Cognitive neuroscience of emotional memory. *Nature Reviews Neuroscience*, *7*, 54–64.

12   Talmi, D., Anderson, A. K., Riggs, L., Caplan, J. B., & Moscovitch, M. (2008). Immediate memory consequences of the effect of emotion on attention to pictures. *Learning & Memory*, *15*, 172–182. https://doi.org/10.1101/lm.722908

13   Blanchette, I., & Richards, A. (2010). The influence of affect on higher level cognition: A review of research on interpretation, judgement, decision making and reasoning. *Cognition & Emotion*, *24*, 561–595. doi:10.1080/02699930903132496.

14   Lerner, J. S., & Keltner, D. (2000). Beyond valence: Toward a model of emotion-specific influences on judgement and choice. *Cognition & Emotion*, *14*, 473–493. https://doi.org/10.1080/026999300402763

15   Priedols, M., & Dimdins, G. (2022). Confirmation bias, analytical thinking, and emotional intensity in evaluating news headlines online. *Human, Technologies and Quality of Education*, 39–52. https://doi.org/10.22364/htqe.2022.03

16   Williams, J. L. (2009). *Emotional framing effects*. University of Minnesota. http://purl.umn.edu/52608

17   Chu, B., Marwaha, K., Sanvictores, T., Awosika, A. O., & Ayers, D. (2024). *Physiology, stress reaction*. In: *StatPearls*. StatPearls Publishing.

18   West, M. (2021, July 29). What is the fight, flight, or freeze response? *MedicalNewsToday*. www.medicalnewstoday.com/articles/fight-flight-or-freeze-response

19   Salleh, M. R. (2008). Life event, stress and illness. *Malaysian Journal of Medical Sciences*, *15*(4), 9–18.

20   American Psychological Association. (2023, March 8). Stress effects on the body. *Psychology Topics*. www.apa.org/topics/stress/body

21   Klein, G. (1996). The effect of acute stressors on decision making. In J. E., Driskell & E. Salas (Eds.), *Stress and human performance*. Psychology Press, pp. 49–88.

22   McLaughlin, K. A., Borkovec, T. D., & Sibrava, N. J. (2007). The effects of worry and rumination on affect states and cognitive activity. *Behavior Therapy*, *38*, 23–38. https://doi.org/10.1016/j.beth.2006.03.003

23   Kabat-Zinn, J. (2003). Mindfulness-based stress reduction (MBSR). *Constructivism in the Human Sciences*, *8*, 73–107.

24   Zeidan, F., Johnson, S. K., Diamond, B. J., David, Z., & Goolkasian, P. (2010). Mindfulness meditation improves cognition: Evidence of brief mental training, *Consciousness and Cognition*, *19*, 597–605.

25   MacLean, K. A., Ferrer, E., Aichele, S. R., Bridwell, D. A., Zanesco, A. P., Jacobs, T. L., King, B. G., Rosenberg, E. L., Sahdra, B. K., Shaver, P. R., Wallace, B. A., Mangun, G. R., & Saron, C. D. (2010). Intensive meditation training improves perceptual discrimination and sustained attention. *Psychological Science*, *21*, 829–839. https://doi.org/10.1177/0956797610371339

26   Hölzel, B. K., Carmody, J., Vangel, M., Congleton, C., Yerramsetti, S. M., Gard, T., & Lazar, S. W. (2011). Mindfulness practice leads to increases in regional brain gray matter density. *Psychiatry Research: Neuroimaging, 191*, 36–43. https://doi: 10.1016/j.pscychresns.2010.08.006

27   Purdon, C. (2021). Cognitive restructuring. In A. Wenzel (Ed.), *Handbook of cognitive behavioral therapy: Overview and approaches*. American Psychological Association, pp. 207–234. https://doi.org/10.1037/0000 218-008

28   Traeger, L. (2013). Cognitive restructuring. In M. D. Gellman, & J. R. Turner (Eds.), *Encyclopedia of behavioral medicine*. Springer. https://doi.org/ 10.1007/978-1-4419-1005-9_166

29   Ezawa, I. D., & Hollon, S. D. (2023). Cognitive restructuring and psycho-therapy outcome: A meta-analytic review. *Psychotherapy, 60*, 396–406. https://doi.org/10.1037/pst0000474

30   Li, J. Z., Herderich, A., & Goldenberg, A. (2024, April 19). Skill but not effort drives GPT overperformance over humans in cognitive reframing of negative scenarios. *PsyArXiv Preprints*. https://osf.io/preprints/psyarxiv/fzvd8

31   Eliot, L. (2024, May 3). Generative AI and the great promise of single-session therapy for mental health. *Forbes Newsletters*. www.forbes.com/sites/lanceel iot/2024/05/03/generative-ai-and-the-great-promise-of-single-session-ther apy-for-mental-health/?sh=682d80552ec6

32   Loboda, K. (2023, August 3). Revolutionizing mental health: Generative AI in therapy. *Productive/Edge*. www.productiveedge.com/blog/revolutioniz ing-mental-health-generative-ai-and-therapy

33   Torous, J., & Blease, C. (2024). Generative artificial intelligence in mental health care: Potential benefits and current challenges. *World Psychiatry, 23*, 1–2. doi:10.1002/wps.21148.

34   Schwarz, N., & Clore, G. L. (1983). Mood, misattribution, and judgments of well-being: Informative and directive functions of affective states. *Journal of Personality and Social Psychology, 45*, 513–523. https://doi.org/10.1037/ 0022-3514.45.3.513

35   Bower, G. H. (1981). Mood and memory. *American Psychologist, 36*, 129–148. https://doi.org/10.1037/0003-066X.36.2.129

36   Rigoni, D., Demanet, J., & Sartori, G. (2015). Happiness in action: The impact of positive affect on the time of the conscious intention to act. *Frontiers in Psychology, 6*. www.frontiersin.org/journals/psychology/articles/10.3389/ fpsyg.2015.01307

37   Renner, F., Schwarz, P., Peters, M. L., & Huibers, M. J. H. (2014). Effects of a best-possible-self mental imagery exercise on mood and dysfunctional attitudes. *Psychiatry Research, 215*, 105–110. https://doi: 10.1016/j.psychres.2013.10.033

38   O'Donnell, C., Di Simplicio, M., Brown, R., Holmes, E. A., & Heyes, S. B. (2018). The role of mental imagery in mood amplification: An investigation across subclinical features of bipolar disorders. *Cortex, 105*, 104–117.

39   Garland, E. L., Farb, N. A., Goldin, P. R., & Fredrickson, B. L. (2015). Mindfulness broadens awareness and builds eudaimonic meaning: A process

model of mindful positive emotion regulation. *Psychological Inquiry, 26,* 293–314. doi:10.1080/1047840X.2015.1064294.

40   Zeidan, F., Johnson, S. K., Diamond, B. J., David, Z., & Goolkasian, P. (2010). Mindfulness meditation improves cognition: Evidence of brief mental training. *Consciousness and Cognition: An International Journal, 19,* 597–605. https://doi.org/10.1016/j.concog.2010.03.014

41   Damasio, A. (2005). *Descartes' error: Emotion, reason, and the human brain.* Penguin.

42   Beijamini, F., Valentin, A., Jäger, R., Born, J., & Diekelmann, S. (2021). Sleep facilitates problem solving with no additional gain through targeted memory reactivation. *Frontiers in Behavioral Neuroscience, 15,* 645110. doi:10.3389/fnbeh.2021.645110.

43   Sio, U. N., Monaghan, P., & Ormerod, T. (2013). Sleep on it, but only if it is difficult: Effects of sleep on problem solving. *Memory and Cognition, 41,* 159–166. doi:10.3758/s13421-012-0256-7.

Part II

Interacting with Humans in the Age of AI

Chapter 6

Navigating Human Personalities

The skills and knowledge we have considered in previous chapters are only a small part of what we need to learn as AIs permeate our world. In this chapter we consider skills and knowledge that we needed well before the advent of AI, but which will become increasingly important in the Age of AI. Specifically, to flourish in the Age of AI, we need to interact well with other people. As machines do more of our work, we will have more time to interact with others. Many of those interactions may hinge on working closely with others to do what we humans do better than AIs—responding in open-ended situations that require taking context into account.

One of the reasons interacting with others can be a challenge is that we need to know how to work with people who have different personalities. "Personality" consists of a person's individual patterns of behavior and emotional responses. These dispositions are influenced by a person's temperament and experiences. People are different, and their differences affect how they respond in specific circumstances.

Researchers have developed many different ways to assess personality. Most of these methods focus on measuring personality *traits,* which are specific aspects of personality that are relatively stable over time and place. However, as we shall see, each trait may express itself differently depending both on what other traits we have and on the particular situation. But before we get to how the pieces interact with each other, we need to get clear on the basics.

The Big Five

When researchers have correlated the results from many personality tests, they discovered that the tests all draw on the same five underlying traits. These five traits are broad categories, which are also called "factors" or "dimensions." These so-called "Big Five" personality traits are widely accepted among psychologists. These five factors are: Openness,

DOI: 10.4324/9781032686653-8

Conscientiousness, Extraversion, Agreeableness, and Neuroticism. You can remember them by using the acronym OCEAN.[1,2,3]

Let's consider each factor in turn. In order to define and illustrate the dimensions, we focus on the implications of high vs. low scores, but keep in mind that people can have intermediate scores.

Openness to Experience

People who score high on the Openness trait tend to be imaginative, curious, open-minded, adventurous, and are sensitive to beauty. In contrast, people who have low scores on this factor tend to prefer familiar routines, not be interested in theories or abstract ideas, and prefer traditional ways of doing things. Researchers have found that scores on tests that assess this dimension predict how well people perform tasks that require creative thinking. In particular, people who score highly on this factor are more productive when they engage in divergent thinking (as we discussed in Chapter 4). Studies have also documented that people who score high on this factor are inclined to engage in artistic and cultural activities.[4,5,6]

But scoring high on this dimension also has its downsides.[7] For example, people who score high may struggle with routine tasks, have unstable goals, sometimes "overthink" a situation and become indecisive, take unreasonable risks, be impulsive, have trouble sticking with a career—or even a partner—and may be prone to experimenting with illegal substances.

In spite of the fact that this factor regularly emerges from analyses of scores on personality tests, it is difficult to assess rigorously and accurately because it encompasses so many different types of behaviors and preferences. This factor is typically measured by asking people to fill out questionnaires, and hence the scores can reflect biases and inaccuracies in a person's self-perception. The scores can also reflect the respondent's desire to present themselves in a positive light. In addition, the concept of "openness" is not entirely distinct from intelligence and creativity, and hence it is difficult to sort out the impact of this personality trait from the contributions of other psychological factors.

Conscientiousness

People who score high on Conscientiousness tend to be well organized, dependable, and have a lot of self-discipline. People who score low on this factor may be more spontaneous, not detail-oriented, and not very goal-oriented. Scores on tests of conscientiousness predict whether a person will follow through on plans, be careful and diligent, and follow rules. Scores on measures of this factor also predict job performance in a wide range of fields. Moreover, scores on measures of this factor predict how long people will live, which may simply reflect the fact that these scores predict behaviors such as regular exercise, healthy eating, and following medical advice.[8,9,10] Indeed, measures of conscientiousness in childhood predict both career achievement and the quality of social relationships later in life. Conscientiousness is also associated with being persistent, which may explain some of these effects.

The potential downsides of being highly conscientious are inflexibility, being obsessed by work to the exclusion of other aspects of life, and being prone to work-related stress. Highly conscientious people may also be slower to respond to new opportunities. However, keep in mind that someone can be both high on Conscientiousness and Openness, which paints a more complex picture of their personality.

Extraversion

This dimension specifies "Extraversion" on one end and "Introversion" on the other. Extraversion refers to being outgoing, liking to talk to and socialize with others, being comfortable expressing emotions, being assertive, and generally being oriented toward other people. In contrast, Introversion refers to being introspective, liking to spend time alone, preferring quiet activities, and generally preferring to interact with one or two other people at a time. People who are extraverted gain energy from interacting with others, whereas people who are introverted gain energy from spending time alone.[11,12,13]

People who score high on this dimension (i.e., toward the Extraversion pole) generally are more positive in their outlook on life and report that they often have positive emotions. They also tend to succeed in jobs that require frequent interactions with other people. Leaders are often extraverted. On the other hand, people who score toward the introversion pole of this dimension can concentrate more deeply, work well independently, listen well, and are keen observers—all of which tends to make them good decision makers.

Cultures vary in the degree to which they value one or the other pole of this dimension. Some cultures, such as the US, primarily value outgoing, sociable extraverts, whereas other cultures, such as many in Asia, primarily value thoughtful, contained introverts. Moreover, even in cultures like the US, some people may be regarded as "too extraverted," engaging in impulsive behaviors and taking risks—which can lead to accidents or substance abuse.

Agreeableness

The Agreeableness factor reflects the degree to which a person is friendly, compassionate, and cooperative.[14,15,16,17] People who score high on this dimension tend to value getting along with others and are often considered good-natured, helpful, and trusting. In contrast, people who score low on this dimension may be more competitive, assertive, and less inclined to cooperate with others. High scores on this factor predict pro-social behavior and positive social outcomes, such as peer acceptance and stable interpersonal relationships. High scores on this factor also predict lower levels of criminal behavior and substance abuse. In addition, high scores predict better conflict resolution skills and fewer interpersonal conflicts.

Nevertheless, scoring high on this factor may be a mixed blessing in some contexts. For one, people who score high on it may not do well when they must negotiate in difficult situations. For another, the cooperative and trusting nature of people who score high on this factor can make them susceptible to being manipulated by others.

Neuroticism

The Neuroticism factor focuses on "emotional reactivity." People who score high on this factor tend to be emotionally unstable and may have extreme negative emotions—such as anxiety, anger, or depression—especially when they are stressed. People who score low tend to be emotionally stable, resilient, and are usually composed. High scores on this factor predict how likely someone is to have mood disorders or anxiety disorders, as well as to have poor health. People who score high on this dimension also are prone to getting stuck in maladaptive thought processes, such as obsessively thinking about a potential problem, dwelling on fears, and inappropriately making molehills into catastrophic mountains, seeing disasters even when they are not likely. These sorts of thought processes can in turn amplify the effects of stress and create other mental health issues.[18,19,20]

Some have criticized this dimension because it often is cast as entirely negative, but some aspects of high neuroticism, such as vigilance and prudence, can be adaptive in high-risk situations.

To make all of this concrete, you might try the following prompt—or you might take a widely used free test at https://bigfive-test.com/; a quick Google search reveals many possibilities. The following prompt does not produce a scientifically valid assessment, but rather is intended just to give you a better sense of the nature of these factors. You can answer for yourself or for someone you know well enough to have insight into their internal experiences. In the below, I responded based on my knowledge of a very close friend.

Me:

Please give me a short test of the Big Five personality dimensions. For each dimension, ask me 4 questions. Ask me to use a five point scale to respond to the questions, where "1" indicates that I completely disagree and "5" indicates that I completely agree, with intermediate scores indicating intermediate levels of agreement. Please pause after you ask me each question, and wait for me to respond. Do not move on until I've responded. After you ask me all the questions, please score my responses. Thank you.

AI:

Great! Let's begin with the first dimension, Openness to Experience. Here are your questions:

1. I have a vivid imagination.
2. I enjoy thinking about abstract concepts.
3. I am open to trying new foods, music, and cultures.
4. I prefer variety to routine.

Please respond to each question using the 1–5 scale. You can list all your answers at once or one by one, whatever you prefer.

Me:

5, 5, 5, 5

AI:

Thank you for your responses on Openness to Experience. Now, let's move on to the next dimension, Conscientiousness. Here are your questions:

1. I like to make plans and follow through with them.
2. I pay attention to details.
3. I get chores done right away rather than procrastinating.
4. I believe that rules and regulations should be followed.

Please rate each statement from 1 (completely disagree) to 5 (completely agree).

Me:

4, 5, 5, 4

AI:

Thank you for your responses on Conscientiousness. Let's continue with the third dimension, Extraversion. Here are your questions:

1. I feel energized when I am around people.
2. I enjoy being the center of attention.
3. I start conversations with people I don't know.
4. I talk to a lot of different people at parties.

Please respond to each question using the 1–5 scale.

Me:

2, 1, 2, 3

AI:

Thanks for your responses on Extraversion. Now, let's move to the fourth dimension, Agreeableness. Here are your questions:

1. I am interested in people's problems.
2. I make people feel at ease.
3. I sympathize with others' feelings.
4. I trust people.

Please rate each statement from 1 (completely disagree) to 5 (completely agree).

Me:

4, 3, 5, 2

AI:

Thank you for your responses on Agreeableness. Lastly, we'll address the dimension of Neuroticism. Here are your questions:

1. I get upset easily.
2. I worry about things.
3. I often feel sad.
4. I am easily stressed out.

Please respond to each question using the 1–5 scale.

Me:

2, 4, 3, 3

AI:

Thank you for completing all the questions! Now let's score your responses for each dimension of the Big Five personality traits:

Openness to Experience: Your scores were 5, 5, 5, 5. Average = (5 + 5 + 5 + 5) / 4 = 5.0

- You are extremely open to new experiences, curious, and imaginative.

Conscientiousness: Your scores were 4, 5, 5, 4. Average = (4 + 5 + 5 + 4) / 4 = 4.5

- You are highly conscientious, reliable, and organized.

Extraversion: Your scores were 2, 1, 2, 3. Average = (2 + 1 + 2 + 3) / 4 = 2.0

- You tend to be introverted, preferring smaller groups or more intimate settings.

Agreeableness: Your scores were 4, 3, 5, 2. Average = (4 + 3 + 5 + 2) / 4 = 3.5

- You are generally cooperative and empathetic, but you may be cautious in your trust towards others.

Neuroticism: Your scores were 2, 4, 3, 3. Average = (2 + 4 + 3 + 3) / 4 = 3.0

- Your level of emotional stability is moderate, with some susceptibility to stress and worry, but not overwhelmingly so.

This overview provides insights into your personality, highlighting areas where you might excel and aspects of yourself you might be more aware of or work on. Remember, personality can be complex and multifaceted, and these scores are just one way to understand individual differences!

As this sort of exercise makes clear, personality is multidimensional and each of us has different dispositions that can come to the fore in different situations.

Implications of the Big Five

An enormous amount of research has supported the existence and importance of these five personality factors. For example, in spite of small variations, researchers usually find very similar results across many different cultures.[21,22,23] Moreover, as noted earlier, scores on measures of the Big Five traits predict how well people succeed in many aspects of life, ranging from how well they perform their jobs, to how happy they are with their relationships, to how healthy they are.[24,25] In addition, when researchers have examined identical vs. fraternal twins, who share 100% and 50% of their genes, respectively, they consistently find greater similarity in the Big Five scores for identical twins.[26,27] This finding suggests that genetics influence the factors. In fact, researchers have estimated that somewhere between 40% and 60% of personality is determined by genes. Consistent with the role of genetics in personality, studies have also shown that the Big Five traits are relatively stable over time.[28,29,30] Although changes can occur as we age—because we have had powerful experiences or tried hard to change—for the most part our personalities are stable over time.[31]

Traits in Context

The impact of one trait typically depends both on our other traits and on the specific situation. For example, particular combinations of factors affect how well a given person fits in a given work environment. The exact same combination of traits can help or hinder, depending on the context. For instance:

- *High Openness + High Conscientiousness.* A person who has this particular combination of traits should do well when innovation is encouraged, such as occurs in tech startups. They not only contribute new ideas, but also are disciplined enough to be able to implement them. But what is a virtue in some types of work is a drawback in others. For example, the same person might feel smothered in a highly bureaucratic company or organization, and might experience stress as they remain open to new ideas while adhering to established rules and procedures.
- *High Extraversion + Low Agreeableness.* Someone with this combination of traits would probably thrive in competitive sales or leadership roles. These sorts of roles often require being assertive and having a thick skin. But by the same token, this same sort of person would encounter problems in a work environment where teamwork and harmony were paramount, such as occurs in some non-governmental organizations.

Similarly, different combinations of traits have clear implications for how well a person will do in different educational settings. For example:

- *High Conscientiousness + Low Openness.* People who have this combination of traits do well in traditional educational settings, which have clear rules, expectations, and feedback. These same people have a harder time in less structured educational programs, such as those that emphasize creative research or hinge on asking good questions.
- *Low Conscientiousness + High Extraversion.* These people would do well in programs that emphasize group work and the value of public presentations. However, they might not do well in programs that hinge on individual work and long-term projects that require prolonged attention and diligence.

By the same token, different combinations of traits affect how people interact in specific social contexts. Examples:

- *High Neuroticism + High Agreeableness.* People with this particular combination of traits tend to be sensitive and responsive partners in an intimate relationship. However, they might also have a lot of anxiety and could become cloying and insincere.

• *High Extraversion + Low Neuroticism.* Such people would do well in most social situations. They are usually resilient to social setbacks, such as rejection, and don't take negative feedback personally. However, such people might have difficulty maintaining an intimate discussion that requires being emotionally responsive and vulnerable.

These examples illustrate that if we want to understand the impact of personality differences on work and life, we need to know more than how a person scores on each of the five individual factors. The combinations matter. But more than that, the effects of a given combination depend on the specific context.

It's worth noting that AIs will have difficulty responding to the complex interplay among personality traits in different contexts, especially in open-ended situations where unpredicted factors may come into play (see Chapter 1). This is the case if only because there is an open-ended (near infinite?) number of combinations of traits and contexts, and AIs cannot be trained on all of them in advance. Nor can we be sure that AIs have abstracted out the relevant underlying dimensions from their training materials, which presumably would allow them to generalize to at least some new situations. We humans have an edge over AIs when dealing with the nuances of personalities in context, and are likely to continue to have one for the foreseeable future.

Beyond the Big Five

We have been focused at a high level, considering each of the Big Five personality factors. However, each of these factors is in fact a broad category, subsuming numerous more specific traits. Let's now dive deeper, and take a closer look at more specific traits. This is well worth doing because these more specific traits typically are better predictors of behavior than the broad categories themselves.[32,33,34]

Identifying Central Traits

Table 6.1 presents a summary of ten constituent traits for each of the Big Five factors. In spite of the length this list, this is only a sample of the large number of such traits researchers have identified. These traits are powerful predictors of behavior in part because only some of them are relevant for any given individual.[35] Our *central traits* are those that are usually in the foreground, affecting our thoughts, feelings and behavior across a wide range of situations, whereas more peripheral traits may only be important in particular situations. For example, "Self-Efficacy," one of the constituent

Table 6.1 Examples of Constituent Traits of the Big Five.

Big Five Factor	Individual Traits	Explanation of Traits	Example of Traits
Openness	Imagination	Tendency to think about and explore abstract concepts and ideas.	Conceiving of an alternative history where World War I was never fought.
	Artistic Interests	Appreciation of art and beauty, both in nature and in culture.	Enjoying visits to art galleries and museums.
	Emotionality	Awareness and expression of one's own feelings.	Easily moved by emotional stories in movies.
	Adventurousness	Willingness to try new activities and experience new things.	Trying exotic foods and exploring unfamiliar places.
	Intellect	Engagement with new ideas and love of learning.	Enjoying puzzles and reading about new scientific discoveries.
	Liberalism	Readiness to challenge authority, convention, and traditional values.	Advocating for progressive policies in society.
	Curiosity	A strong desire to learn or know about anything.	Asking detailed questions about how things work.
	Creativity	Ability to produce novel solutions or products.	Coming up with a unique design for a product.
	Unconventionality	Preference for unique and unconventional aesthetics or behaviors.	Wearing eclectic clothes that stand out.
	Openness to Emotions	Readiness to embrace complex and multifaceted emotional experiences.	Reflecting deeply on varied emotional states.
Conscientiousness	Self-Efficacy	Belief in one's own competence to complete tasks and reach goals.	Setting and achieving challenging career goals.
	Orderliness	Preference for organization, routine, and a structured approach to life.	Keeping a detailed planner and a tidy workspace.

(Continued)

Table 6.1 (Continued)

Big Five Factor	Individual Traits	Explanation of Traits	Example of Traits
	Dutifulness	Sense of duty and obligation to fulfill moral and contractual commitments.	Always fulfilling promises and meeting deadlines.
	Achievement-Striving	Effort put towards setting and accomplishing challenging goals.	Working overtime to excel in a career.
	Self-Discipline	Ability to start tasks and follow through to completion despite distractions or boredom.	Consistently sticking to a fitness routine.
	Cautiousness	Tendency to think through possibilities before acting or speaking.	Reviewing all potential outcomes before making a decision.
	Prudence	Making wise decisions based on foresight and practicality.	Avoiding risky investments and planning for the future.
	Perfectionism	Desire to achieve flawlessness and set high performance standards.	Double-checking work for errors to ensure perfection.
	Responsibility	Accepting accountability for one's actions and being dependable.	Being a reliable point of contact in critical situations.
	Persistence	Steadfastness in doing something despite difficulty or delay in achieving success.	Continuing to work on a challenging problem until solved.
Extraversion	Friendliness	Outgoing and sociable behavior, enjoying being with people.	Always greeting people warmly and starting conversations.
	Gregariousness	Preference for being in the company of others rather than alone.	Hosting parties and social gatherings.
	Assertiveness	Tendency to take charge and direct the activities of others.	Leading group projects or meetings confidently.

Table 6.1 (Continued)

Big Five Factor	Individual Traits	Explanation of Traits	Example of Traits
	Activity Level	Enjoyment of keeping busy and engaged.	Participating in multiple hobbies and sports.
	Excitement-Seeking	Seeking stimulation in the company of others or in activities.	Going bungee jumping or enjoying lively social events.
	Cheerfulness	Presence of positive emotions and outlook on life.	Often smiling and maintaining a positive demeanor.
	Warmth	Approachability and easy rapport with others.	Making new acquaintances feel at ease.
	Positivity	Tendency to see the best in situations and expect good outcomes.	Always speaking about the bright side of things.
	Energy	High level of enthusiasm and vitality.	Always energetic and ready for new adventures.
	Talkativeness	Inclination to engage in conversation.	Dominating conversations at social gatherings.
Agreeableness	Trust	Belief in the sincerity and good intentions of others.	Assuming people have the best intentions.
	Morality	Simplicity in being frank, genuine, and sincere.	Being straightforward and honest without being calculating.
	Altruism	Enjoyment of helping others; selflessness.	Volunteering regularly at local shelters.
	Cooperation	Dislike of confrontations. Preference for getting along with others.	Avoiding arguments and seeking compromise.
	Modesty	Tendency to play down own achievements or abilities.	Rarely bragging about personal accomplishments.
	Sympathy	Feeling compassion towards others in distress.	Feeling sad for a friend's problems and trying to help.

(Continued)

Table 6.1 (Continued)

Big Five Factor	Individual Traits	Explanation of Traits	Example of Traits
	Kindness	Being friendly, generous, and considerate.	Helping a neighbor with daily chores without being asked.
	Gentleness	Mild manner and calm disposition.	Speaking softly and treating others with care.
	Flexibility	Willingness to adapt or compromise in the face of differing opinions.	Changing plans to accommodate others' needs.
	Patience	Ability to accept or tolerate delays, problems, or suffering without becoming annoyed.	Waiting calmly for their turn without complaining.
Neuroticism	Anxiety	Tendency to be nervous or easily stressed.	Worrying about minor issues and future possibilities.
	Anger	Frequency and intensity of feelings of anger and frustration.	Getting irritated by small inconveniences.
	Depression	Tendency to feel sad, dejected, and discouraged.	Feeling down after setbacks and having trouble recovering.
	Self-Consciousness	Sensitivity to what others think about oneself.	Frequently wondering how others perceive them.
	Immoderation	Tendency to feel strong cravings and urges.	Struggling to resist overindulging in favorite foods.
	Vulnerability	Difficulty coping with stress or pressure.	Feeling overwhelmed in high-pressure situations.
	Sensitivity	Quick to detect or respond to slight changes, signals, or influences.	Easily affected by slight changes in mood or tone.
	Emotional Volatility	Rapid, intense, and uncontrollable swings in feelings.	Reacting intensely to everyday frustrations.
	Pessimism	Expectation that the worst will happen.	Doubting outcomes even when odds are favorable.
	Insecurity	Lack of confidence in one's own abilities or worth.	Hesitating to take up new challenges due to self-doubt.

traits that underlies the Conscientiousness factor, may be a defining char-acteristic for Sally, but simply not be important for John, one way or the other—and hence it would be considered a central trait for her but not for him. In this case, Sally's score on measures of Self-Efficacy probably pre-dicts her behavior, but John's score probably does not.

Given the predictive power of central traits, it behooves us to become sensitive to the specific traits that underlie the Big Five factors. Table 6.1 is daunting, which usually signals an opportunity to engage the Cognitive Amplifier Loop (CAL) to offload the cognitive burden of trying to mem-orize and learn how to apply all this information. However, in this case we need to acknowledge that AIs are not well suited to diagnosing interactions among traits when taking context into account. Nevertheless, the CAL can help us identify which central traits a person may be expressing because these traits are typically evident across different situations.

To get a sense of the central traits being expressed by a person in a given situation, try the following prompt:

> Please ask the user to describe a scenario in which a person did something or said something. Wait until the user has described the scenario; do not continue until the user has completed this descrip-tion. As soon as they have finished, prepare to consult the attached Table to infer the central traits that the person in the scenario is expressing. If the situation is ambiguous, please ask the user for add-itional information to clarify. For example, if more than one person is described in the scenario, ask which one (or ones) the user wants you to focus on. Once you can draw an inference about which trait or traits is clearly being revealed, please explain this to the user. [ATTACH TABLE 6.1]

Here is the response:

> **AI:**
>
> Please describe a scenario in which a person did something or said something. Once you provide the description, I'll consult the attached table to infer the central traits that the person is express-ing. If the situation is ambiguous or involves multiple people, let me know which individual you'd like me to focus on.

Me:

I observed the following interaction: It was a cocktail party, and a young woman from Poland was chatting. She said something, and the guy she was talking to made a joke. Admittedly, it wasn't very funny, but she froze up and said "Hah. Hah. Hah." And turned her back on him. He then apologized profusely for his bad joke, but she was very stiff and unforgiving, barely acknowledging his apology. Given that they didn't know each other, it couldn't have been "personal" in the sense that she didn't like other things about him. What do you think was going on?

AI:

Based on the scenario you described, the young woman's response at the cocktail party reflects several personality traits that can be analyzed using the Big Five personality dimensions. Here's a breakdown:

1. Neuroticism: This dimension often involves emotional sensitivity and volatility. The woman's stiff and unforgiving reaction to the joke, despite the apology, suggests a higher level of Sensitivity and possibly Vulnerability. These traits are characterized by a heightened reaction to perceived slights or social missteps, and difficulty in managing stress or emotional disturbances in social interactions.

2. Agreeableness: This trait involves aspects like trust, cooperation, and sympathy. The woman's behavior might indicate a lower level of Cooperation and Flexibility, as her response was somewhat rigid and she did not seem inclined to smooth over the social interaction after the awkward joke. Instead of trying to maintain a pleasant interaction, she turned away, showing a lack of adaptability in the face of social discomfort.

3. Extraversion: While extraversion typically includes traits like gregariousness and friendliness, her reaction might suggest lower levels of these traits. Her inability to engage in light-hearted banter or brush off the joke could reflect less comfort or desire to engage openly in social settings, which are part of the Cheerfulness and Warmth traits of extraversion.

The scenario points mainly towards elements of Neuroticism (especially sensitivity and vulnerability to emotional stimuli in social settings) and lower Agreeableness concerning flexibility and cooperation. This interpretation suggests she might have a personality that is more sensitive to social cues and less forgiving of what she perceives as social errors, making it difficult for her to dismiss the awkwardness of the situation lightly.

The AI's observations and hypotheses are a helpful way to begin to think about the social interaction. However, we need to keep in mind that the constituent traits may interact in complex ways with the situation, and hence that AI's interpretations are probably too simple. But even so, this sort of response can point us in potentially interesting directions.

Another Type of Traits

We have focused on the Big Five model and its constituent traits because this is the most accepted and validated approach to analyzing human personality. However, it's not the only one. It's worth briefly taking a look at one other widely used approach, namely the Myers-Briggs Type Indicator (MBTI).

First, what is the MBTI? It is a questionnaire that is designed to assess four key aspects of personality.[36,37] Katharine Cook Briggs and her daughter Isabel Briggs Myers developed this approach, based on aspects of theories published by the Swiss psychoanalyst Carl Jung. Jung posited that we humans rely on four functions when we interact with the world, namely sensation, intuition, feeling, and thinking. The MBTI uses these functions as a springboard for developing four dichotomous scales, and classifies each person into one of 16 personality types based on the combination of scores on these dimensions. The four dichotomies are as follows:

- Extraversion (E) vs. Introversion (I): This dimension attempts to capture how a person gets or focuses energy, either from interacting with people and engaging in social activities or from reflection and solitary activities. This is the same as the Extraversion factor of the Big Five.
- Sensing (S) vs. Intuition (N): This dimension attempts to capture how a person gathers information, either via the senses (Sensing) or by noting patterns and meanings beyond the immediate data (Intuition).
- Thinking (T) vs. Feeling (F): This dimension attempts to capture how a person makes decisions, either by relying on objective facts and

principles (Thinking) or by relying on values and emotional impact (Feeling).

- Judging (J) vs. Perceiving (P): This dimension attempts to capture how a person relates to the outside world, either by preferring to plan and structure (Judging) or by preferring to be open, adaptable, and flexible (Perceiving).

Combinations of these four preferences produce 16 personality types, each of which is specified by a four-letter code (e.g., ENTP, ISTP).

Although the MBTI has intuitive appeal to many, scientific studies have repeatedly failed to support it.[38,39] Indeed, up to 50% of people who take this test will get different results for at least one of the dimensions when they take it again five weeks later. This is very poor reliability.[40,41] But more than that, the test fails to predict behavior in work performance, relationships, or personal satisfaction.[42] Perhaps an even deeper problem is that there are only 16 categories, but personality differences are more varied and nuanced—as Table 6.1 attests. Finally, to the extent there is any utility to the MBTI, it appears to result from correlations with specific Big Five factors. In particular, the Extraversion vs. Introversion scale and the Sensing vs. Intuition scales are highly correlated with the Extraversion and Agreeableness scales, respectively.[43]

Changing Traits, Changing Behavior

Our personality traits are relatively stable over time, but they are not absolutely fixed. Researchers have found that various types of interventions can change traits. As a salient example, let's look at just one Big Five factor, Conscientiousness. Researchers have long known that measures of conscientiousness and intelligence are excellent predictors of success on the job and in life.[44,45,46] Indeed, these measures predict life outcomes as varied as divorce, occupational attainment, and even how long someone will live. Thus, it's worth a closer look at this factor.

Various sorts of events can change this trait (e.g.,[47,48]). Such events include the following:

- *Life Experiences.* Some types of major changes in our personal situations, such as getting married or having children, may increase conscientiousness.
- *Aging.* People generally become more conscientious over time, which probably reflects a combination of maturational changes in the brain and accumulated life experiences.

- *Effort.* We can also increase our conscientiousness deliberately, in part by setting goals and ensuring that we follow through on them. This in turn can lead to new habits, which can make us even more conscientious.
- *Environment.* Switching to a new job that is highly structured can make us more conscientious. For example, managing a fast-food outlet, with its set daily tasks, can lead us to develop an eye for detail and structure.

Although we can change our personalities to some degree, this generally requires a lot of effort. Another strategy is not to try to change ourselves, but rather to accept our personalities and figure out how to adapt our *behavior* to specific contexts. To get started, take a personality test that assesses your particular traits on the Big Five (such as the one cited earlier). You can then seek feedback from trusted friends, family, or colleagues about the accuracy of your assigned traits. Such feedback can be surprising. We may not closely monitor how our behaviors reflect what we think and feel, and we may not be aware of how others perceive those behaviors in different contexts. As a result of this exercise, we can gain a good sense of how we score on the Big Five factors and how they play out in particular situations.

Once we have a good idea of these personality factors, we then need to be sensitive to situations where they are helping us vs. creating friction and stress. If personality traits are helping us, we should leave well enough alone. But if our personality is getting in the way, we would be better off adjusting our behavior to sidestep the problem. That is, even if we can't easily change the trait itself, we can—with some effort—change our behavior. To get better at doing this, we can use the CAL.

For example, we can instruct an AI to develop role-playing exercises that help us prepare for situations where we need to resist our natural impulses. For instance, if we are introverted—perhaps we have low central traits of Talkativeness and Assertiveness, as noted in Table 6.1—and about to go to a networking event, we can engage with the AI along the lines of what I did below.

Me:

I am about to go to a networking event. My natural tendency is to be introverted and not reach out to people in such situations. Can you please develop a series of social networking scenarios. For each one, first describe the specific scenario and then ask me to role-play what I should do. Pause and wait for my response. Following this, please give me feedback, offering suggestions about what I should have done differently. Then give me a variation of the initial scenario, and repeat the process until I have responded more appropriately. Repeat this entire process until I ask you to stop.

AI:

That sounds like a great exercise to enhance your networking skills. Let's begin with the first scenario:

Scenario 1: The Introduction at a Networking Event
You're at a networking event and you notice a small group of people discussing a topic that interests you. They are casually dressed, seem approachable, and one of them makes eye contact and smiles at you.

Your Task: Approach the group and introduce yourself. What do you say?

(Please respond with your approach, and I'll provide feedback afterward.)

Me:

Hello you all! I couldn't help but overhear what you are talking about. Would you mind if I joined in?

AI:

That's a great start! Your approach is polite and shows genuine interest in the conversation, which is a good way to engage with a group.

Feedback:

Positive Aspects: You started with a friendly greeting and politely asked if you could join, which is respectful and considerate.

Suggestions: You might also want to briefly introduce yourself, including your name and perhaps a line about what brings you to the event or your interest in the topic they are discussing. This can provide a quick context for your presence and interest.

Variation of Scenario 1:

You approach the same group, but this time they seem deeply engaged in their discussion and haven't noticed you yet.

Your Task: How would you approach and introduce yourself in this scenario?

Me:

I would stand at the periphery, and wait patiently for a pause in the conversation and then introduce myself.

AI:

That's a thoughtful and considerate approach. Waiting for a natural pause shows respect for the ongoing conversation and indicates that you're mindful of social dynamics.

Feedback:

Positive Aspects: Waiting for the right moment is a good strategy. It shows that you're attentive and not intrusive.

Suggestions: When you find the appropriate moment to speak up, you could say something like, "Hi, I hope I'm not interrupting, but I found your discussion really interesting. I'm [Your Name], and I'm here because [mention a brief reason or your interest]." This way, you acknowledge the ongoing conversation and introduce yourself smoothly.

New Scenario 2: One-on-One Interaction

You're at the refreshment table and someone who seems to be alone is standing next to you, looking around the room somewhat uncertainly.

Your Task: Initiate a conversation with this person. What do you say?

(After your response, I'll provide feedback.)

[TRUNCATED]

The sorts of tips the AI developed are coping strategies. We can also develop more general coping strategies. For example, even if we score very high on Extraversion and/or Agreeableness, and hence are usually at ease and in our element when we are with other people, we can still improve how we interact with people in specific contexts. In particular, we all can enhance our communication skills. To do so, we can prompt an AI to help us learn to be both more receptive to the messages others send and to express ourselves more fully and clearly (e.g.,[49]). In addition, an AI can help us develop time-management skills. This is particularly useful for people who score low on Conscientiousness.

We can also use an AI to help us select appropriate environments. For example, we can ask an AI to suggest work roles that align with our traits, which not only will help us be more effective and less stressed but also will help our employer, colleagues, and organization by allowing us to play to our strengths. An AI can facilitate the usual process, where over time people tend to select environments that align with their thoughts, feelings, and behavior.[50,51,52] For example, introverted people may select quiet situations where they can read in their spare time whereas extraverted people may join stimulating clubs where they can interact with other people. We can also sometimes alter our environments so that they align with our personality traits. Such alterations could be as simple as shifting the location of a desk to avoid distractions.

Flourishing with the CAL

In sum, we should learn about human personalities for two reasons, both of which will help us to flourish in the Age of AI. One is to help us understand ourselves better, which can allow us to predict our reactions to specific situations in advance—and thereby to know what to ask an AI to help us accomplish. Such assistance can help us become more autonomous and in control. Moreover, to the extent that we can discover jobs that are good fits with our personalities, this will improve our job satisfaction and, presumably, performance.

Another reason for learning about human personalities is to interpret how other people behave and react, which can facilitate our personal relationships—and help us perform many sorts of tasks more effectively.

Our understanding of personality is a crucial ingredient of a type of intelligence that affects how we interact with other people. In the following chapter, we consider what is known as "emotional intelligence," what it is, why it's important, and how we can train it.

Digging Deeper

Below are examples of accessible videos that provide a more in-depth treatment of key aspects of the material reviewed in the chapter. In addition, the below search terms can help the reader locate new videos on the topic.

The Big 5 OCEAN Traits Explained—Personality Quizzes: https://www.youtube.com/watch?v=KCwHV9HCxH0

Trait Theory—History of Personality Psychology: https://www.youtube.com/watch?v=oUgCIvKxbAE

- "Big Five personality traits"
- "OCEAN model of personality"
- "Trait theory in personality psychology"
- "Personality differences in human interactions"
- "Navigating different personality types"
- "Understanding personality for effective communication"
- "Personality and situational context"
- "Personality assessment tools"
- "Genetics and experience in personality development"
- "Personality and cognitive, emotional, and behavioral dispositions"
- "Leveraging personality research for self-understanding"
- "AI and personality analysis"

Jordan Peterson | Big 5 Personality Traits: https://www.youtube.com/watch?v=PgIErTULONc

What Causes Our Personality? Genetics vs. Environment: https://www.youtube.com/watch?v=aew2WRAOk9M

- "Cross-cultural validity of the Big Five personality traits"
- "Big Five personality traits and life outcomes"
- "Genetic and environmental influences on personality"
- "Stability of personality traits over time"
- "Interaction between personality traits and context"
- "Combinations of Big Five traits and work performance"
- "Personality trait combinations and educational success"
- "Big Five traits and relationship dynamics"
- "Context-dependent expression of personality traits"
- "Personality psychology research on trait interactions"
- "Applying the Big Five model in real-world settings"

Myers Briggs (MBTI) Explained—Personality Quiz: https://www.youtube.com/watch?v=2ZF4OM6mrrI

The Problem with the Myers-Briggs Type Indicator: https://www.youtube.com/watch?v=tPgKQO-WX28

- "Myers-Briggs Type Indicator (MBTI) explained"
- "Scientific critique of the MBTI"
- "MBTI reliability and validity issues"
- "Relationship between MBTI and Big Five personality traits"
- "Limitations of correlational evidence in personality research"

Can You Change Your Personality? https://www.youtube.com/watch?v=Dy1i-u_2rQ8

How to Hack Networking | David Burkus | TEDx University of Nevada: https://www.youtube.com/watch?v=xFrqZjIDE44

- "Changing personality traits through interventions"
- "Life experiences and personality development"
- "Age-related changes in personality traits"
- "Deliberate effort and personality change"
- "Environmental influences on personality"
- "Adapting behavior to specific contexts"
- "Stress management techniques for personality types"
- "Developing communication skills based on personality"
- "Time management strategies for different personality traits"
- "Modifying environments to align with personality"
- "Expanding comfort zones and personal growth"
- "Understanding personality for self-regulation and interpersonal interactions"

References

1   Costa, P., & McCrae, R. R. (2002). Personality in adulthood: A Five-Factor Theory perspective. *Management Information Systems Quarterly—MISQ*. doi:10.4324/9780203428412.

2   Goldberg, L. R. (1993). The structure of phenotypic personality traits. *American Psychologist, 48*, 26–34. https://doi.org/10.1037/0003-066X.48.1.26

3   McCrae, R. R., & John, O. P. (1992). An introduction to the Five-Factor model and its applications. *Journal of Personality, 60*, 175–215. doi:10.1111/j.1467-6494.1992.tb00970.x.

4   Kaufman, S. B., Quilty, L. C., Grazioplene, R. G., Hirsh, J. B., Gray, J. R., Peterson, J. B., & DeYoung, C. G. (2016). Openness to experience and intellect differentially predict creative achievement in the arts and sciences. *Journal of Personality, 84*, 248–258. doi:10.1111/jopy.12156.

5   McCrae, R. R., & Sutin, A. R. (2009). Openness to experience. In M. R. Leary & R. H. Hoyle (Eds.), *Handbook of individual differences in social behavior.* The Guilford Press, pp. 257–273.

6   Schretlen, D. J., van der Hulst, E. J., Pearlson, G. D., & Gordon, B. (2010). A neuropsychological study of personality: Trait openness in relation to intelligence, fluency, and executive functioning. *Journal of Clinical and Experimental Neuropsychology, 32,* 1068–1073. doi:10.1080/13803391003689770.

7   Piedmont, R. L., Sherman, M. F., & Sherman, N. C. (2012). Maladaptively high and low openness: The case for experiential permeability. *Journal of Personality, 80,* 1641–1668. doi:10.1111/j.1467-6494.2012.00777.

8   Bogg, T., & Roberts, B. W. (2013). The case for conscientiousness: Evidence and implications for a personality trait marker of health and longevity. *Annals of Behavioral Medicine, 45,* 278–288. doi:10.1007/s12160-012-9454-6.

9   Jackson, J. J., Wood, D., Bogg, T., Walton, K. E., Harms, P. D., & Roberts, B. W. (2010). What do conscientious people do? Development and validation of the Behavioral Indicators of Conscientiousness (BIC). *Journal of Research in Personality, 44,* 501–511. doi:10.1016/j.jrp. 2010.06.005.

10   Roberts, B. W., Jackson, J. J., Fayard, J. V., Edmonds, G., & Meints, J. (2009). Conscientiousness. In M. R. Leary & R. H. Hoyle (Eds.), *Handbook of individual differences in social behavior.* The Guilford Press, pp. 369–381.

11   Haddock, A. D., & Rutkowski, A. P. (Eds.). (2014). *The psychology of extraversion.* Nova Science Publishers.

12   Lucas, R. E., & Diener, E. (2001). Extraversion. In N. J. Smelser & P. B. Baltes (Eds.), *International encyclopedia of the social & behavioral sciences.* Pergamon, pp. 5202–5205.

13   McCabe, K. O., & Fleeson W. (2012). What is extraversion for? Integrating trait and motivational perspectives and identifying the purpose of extraversion. *Psychological Science, 23,* 1498–1505. doi:10.1177/0956797612444904.

14   Furnham, A. (2017). Agreeableness. In V. Zeigler-Hill & T. Shackelford (Eds.), *Encyclopedia of personality and individual differences.* Springer, pp. 1–11. https://doi.org/10.1007/978-3-319-28099-8_1200-1

15   Furnham, A., & Cheng, H. (2015). Early indicators of adult trait agreeableness. *Personality and Individual Differences, 73,* 67–71. https://doi.org/10.1016/j.paid.2014.09.025

16   Graziano, W. G., Habashi, M. M., Sheese, B. E., & Tobin, R. M. (2007). Agreeableness, empathy, and helping: A person x situation perspective. *Journal of Personality and Social Psychology, 93,* 583–599. doi:10.1037/0022-3514.93.4.583.

17   Sheese, B. E., & Graziano, W. G. (2004). Agreeableness. In C. Spielberger (Ed.), *Encyclopedia of applied psychology.* Academic Press, pp. 117–122.

18   Luo, J., Zhang, B., Cao, M., & Roberts, B. W. (2023). The stressful personality: A meta-analytical review of the relation between personality and stress. *Personality and Social Psychology Review, 27,* 128–194. doi:10.1177/10888683221104002.

19   Perkins, A. M., Arnone, D., Smallwood, J., & Mobbs, D. (2015). Thinking too much: Self-generated thought as the engine of neuroticism. *Trends in Cognitive Sciences, 19,* 492–498.

20   Widiger, T. A., & Oltmanns, J. R. (2017). Neuroticism is a fundamental domain of personality with enormous public health implications. *World Psychiatry, 16*, 144–145. doi:10.1002/wps.20411.

21   De Raad, B., Perugini, M., Hrebíčková, M., & Szarota, P. (1998). Lingua franca of personality: Taxonomies and structures based on the psycholexical approach. *Journal of Cross-Cultural Psychology, 29*, 212–232. https://doi.org/10.1177/0022022198291011

22   McCrae, R. R., Terracciano, A., & Members of the Personality Profiles of Cultures Project. (2005). Personality profiles of cultures: Aggregate personality traits. *Journal of Personality and Social Psychology, 89*, 407–425. doi:10.1037/0022-3514.89.3.407.

23   Schmitt, D. P., Allik, J., McCrae, R. R., & Benet-Martínez, V. (2007). The geographic distribution of Big Five personality traits: Patterns and profiles of human self-description across 56 nations. *Journal of Cross-Cultural Psychology, 38*, 173–212. doi:10.1177/0022022106297299.

24   Barrick, M. R., & Mount, M. K. (1991). The Big Five personality dimensions and job performance: A meta-analysis. *Personnel Psychology, 44*, 1–26. https://doi.org/10.1111/j.1744-6570.1991.tb00688.x

25   Roberts, B. W., Kuncel, N. R., Shiner, R., Caspi, A., & Goldberg, L. R. (2007). The power of personality: The comparative validity of personality traits, socioeconomic status, and cognitive ability for predicting important life outcomes. *Perspectives on Psychological Science, 2*, 313–345. doi:10.1111/j.1745-6916.2007.00047.x.

26   Jang, K. L., Livesley, W. J., & Vernon, P. A. (1996). Heritability of the Big Five personality dimensions and their facets: A twin study. *Journal of Personality, 64*, 577–591. doi:10.1111/j.1467-6494.1996.tb00522.x.

27   Loehlin, J. C., & Nichols, R. C. (1976). *Heredity, environment, and personality: A study of 850 sets of twins.* University of Texas Press.

28   Cobb-Clark, D. A., & Schurer, S. (2012). The stability of big-five personality traits. *Economics Letters, 115*, 11–15. https://doi.org/10.1016/j.econlet.2011.11.015

29   Roberts, B. W., & DelVecchio, W. F. (2000). The rank-order consistency of personality traits from childhood to old age: A quantitative review of longitudinal studies. *Psychological Bulletin, 126*, 3–25. doi:10.1037/0033-2909.126.1.3.

30   Soldz, S., & Vaillant, G. E. (1999). The Big Five personality traits and the life course: A 45-year longitudinal study. *Journal of Research in Personality, 33*, 208–232. doi:10.1006/jrpe.1999.2243.

31   Briley, D. A., & Tucker-Drob, E. M. (2014). Genetic and environmental continuity in personality development: A meta-analysis. *Psychological Bulletin, 140*, 1303–1331. https://doi.org/10.1037/a0037091

32   Mershon, B., & Gorsuch, R. L. (1988). Number of factors in the personality sphere: Does increase in factors increase predictability of real-life criteria? *Journal of Personality and Social Psychology, 55*, 675–680. doi:10.1037/0022-3514.55.4.675.

33   Paunonen, S. V., & Ashton, M. C. (2001). Big Five factors and facets and the prediction of behavior. *Journal of Personality & Social Psychology, 81*, 524–539. doi:10.1037/0022-3514.81.3.524.

34   Postigo, Á., Cuesta, M., García-Cueto, E., Prieto-Díez, F., & Muñiz, J. (2021). General versus specific personality traits for predicting entrepreneurship. *Personality and Individual Differences, 182*, Article 111094. https://doi.org/ 10.1016/j.paid.2021.111094

35   Allport, G. W. (1937). *Personality: A psychological interpretation.* Henry Holt.

36   Myers, I. B., & McCaulley, M. H. (1985). *Manual: A guide to the development and use of the Myers-Briggs Type Indicator.* Consulting Psychologists Press.

37   Myers, I. B., McCaulley, M. H., Quenk, N. L., & Hammer, A. L. (1998). *MBTI Manual: A guide to the development and use of the Myers-Briggs Type Indicator* (3rd ed.). Consulting Psychologists Press.

38   Boyle, G. J. (1995). Myers-Briggs Type Indicator (MBTI): Some psychometric limitations. *Australian Psychologist, 30*, 71–74. doi:10.1111/j.1742-9544.1995.tb01750.x.

39   Bradford, V. (2018, October 2). *MBTI facts & common criticisms.* The Myers-Briggs Company. www.themyersbriggs.com/en-US/Connect-With-Us/Blog/ 2018/October/MBTI-Facts--Common-Criticisms

40   Howes, R. J., & Carskadon, T. G. (1979). Test-retest reliabilities of the Myers-Briggs Type Indicator as a function of mood changes. *Research in Psychological Type, 2*, 67–72.

41   Pittenger, D. J. (1993). Measuring the MBTI… and coming up short. *Journal of Career Planning and Placement, 54*, 48–52.

42   Nowack, K. (1996). Is the Myers Briggs Type Indicator the right tool to use? *Performance in Practice, 6.* American Society of Training and Development.

43   McCrae, R. R., & Costa Jr., P. T. (1989). Reinterpreting the Myers-Briggs Type Indicator from the perspective of the Five-Factor Model of Personality. *Journal of Personality, 57,* 17–40. doi:10.1111/j.1467-6494.1989.tb00759.x.

44   Barrick, M. R., & Mount, M. K. (1991). The Big Five personality dimensions and job performance: A meta-analysis. *Personnel Psychology, 44*, 1–26. https://doi.org/10.1111/j.1744-6570.1991.tb00688.x

45   Roberts, B. W., Kuncel, N. R., Shiner, R., Caspi, A., & Goldberg, L. R. (2007). The power of personality: The comparative validity of personality traits, socioeconomic status, and cognitive ability for predicting important life outcomes. *Perspectives on Psychological Science, 2*, 313–345. doi:10.1111/ j.1745-6916.2007.00047.x.

46   Schmidt, F. L., & Hunter, J. E. (1998). The validity and utility of selection methods in personnel psychology: Practical and theoretical implications of 85 years of research findings. *Psychological Bulletin, 124*, 262–274. doi:10.1037/0033-2909.124.2.262.

47   Lüdtke, O., Roberts, B. W., Trautwein, U., & Nagy, G. (2011). A random walk down university avenue: Life paths, life events, and personality trait change at the transition to university life. *Journal of Personality and Social Psychology, 101*, 620–637. doi:10.1037/a0023743.

48   Specht, J., Egloff, B., & Schmukle, S. C. (2011). Stability and change of personality across the life course: The impact of age and major life events on mean-level and rank-order stability of the Big Five. *Journal of Personality and Social Psychology, 101*, 862–882. doi:10.1037/a0024950.

49   Kosslyn, S. M. (2023). *Active learning with AI: A practical guide*. Alinea Learning.

50   Briley, D. A., & Tucker-Drob, E. M. (2014). Genetic and environmental continuity in personality development: A meta-analysis. *Psychological Bulletin, 140*, 1303–1331. https://doi.org/10.1037/a0037091

51   Kendler, K. S., & Baker, J. H. (2007). Genetic influences on measures of the environment: A systematic review. *Psychological Medicine, 37*, 615–626. doi:10.1017/S0033291706009524.

52   Plomin, R., DeFries, J. C., & Loehlin, J. C. (1977). Genotype-environment interaction and correlation in the analysis of human behavior. *Psychological Bulletin, 84*, 309–322. doi:10.1037/0033-2909.84.2.309.

Chapter 7

Enabling Emotional Intelligence

Have you ever noticed that someone you know—a spouse, friend, or colleague—is "better with people" than you are? You've noticed that they are more adept at navigating difficult situations, better able to be on top of their own emotions, better able to gauge other people's states of mind, and perhaps better at knowing who to trust and who to watch closely? Such abilities are all hallmarks of having high *emotional intelligence* (EI). EI is the ability to be aware of and regulate our own emotions as well as to recognize and appropriately respond to other people's emotions. EI is crucial for building deep, satisfying human relationships—which is one key to flourishing, both in the present and in the years to come.

I noted earlier that as AIs move into carrying out increasing numbers of jobs, we humans will have more time to spend with other people, and thus will come to focus increasingly on our personal interactions. If so, EI will become even more important in the Age of AI. Moreover, AIs are not likely to develop high EI in the near future; they are not good at apprehending connections between inferred states of mind and open-ended situations that require taking context into account. Thus, learning to develop and enhance EI is well worth doing.

EI rests on two broad competencies: First, it requires being able to recognize, interpret, and manage our own emotions. Second, it requires being able to recognize, interpret, and influence the emotions of others. In both cases, EI depends on being aware of the nature of emotions, how emotions shape thoughts and actions, and how to regulate emotions. EI is a key factor in personal and professional success, in part because it enhances communication, problem solving, and relationship management.

In what follows we first review major theories of EI, then consider evidence that EI is important, and then discuss how to use the CAL (Cognitive Amplifier Loop) to enhance EI.

DOI: 10.4324/9781032686653-9

Table 7.1 Major Theories of Emotional Intelligence.

Name of Model	Summary of Model	Key Criticisms
Four-Branch Model	This model posits that EI consists of four abilities: 1) perceiving emotions; 2) using emotions to facilitate cognitive processes; 3) understanding emotions; and 4) managing emotions. It emphasizes the role of EI in enhancing cognitive activities, academic performance, and interpersonal interactions.	Critics argue that this model overlaps with personality and general intelligence, questioning whether it taps a distinct ability.
Goleman's EI Model	Goleman's model extends Mayer and Salovey's framework by introducing five main domains of EI: 1) self-awareness; 2) self-regulation; 3) internal motivation; 4) empathy; and 5) social skills. This model is particularly focused on leadership and organizational contexts, suggesting that higher EI leads to better leadership effectiveness and team management.	Criticisms include the conflation of EI with personality traits and a lack of specificity in distinguishing between cognitive and emotional intelligence.
Trait EI Model	This model characterizes EI as a set of personality traits that include adaptability, stress management, emotion regulation, and social awareness. It is based on self-perceived abilities and is assessed through self-report questionnaires. This model links higher trait EI with better coping strategies, life satisfaction, and relationship management.	Critiques focus on the overlap with traditional personality traits such as Extraversion and Neuroticism, and concerns over self-report bias.
Bar-On's Mixed Model	Bar-On's model combines emotional and social skills into a broader model of EI, defining it as a collection of competencies and skills that influence our ability to succeed in coping with environmental demands and pressures. Key components include self-regard, impulse control, and empathy, enhanced by traits like optimism and happiness.	The broad scope of the model leads to criticisms about its overlap with many personality dimensions and reliance on potentially biased self-reports.

What Is Emotional Intelligence?

Different theorists have characterized EI in different ways, as summarized in Table 7.1.

Results from studies provide support for each of these models, which suggests that each of them taps into aspects of EI. Hence, and it's worth considering them all. To begin, the Four-Branch Model of Emotional Intelligence is perhaps the most widely used, and has received the most support.[1,2,3] Research has shown that its measures of EI predict various aspects of both personal and professional relationships (e.g.,[4]). For example, people who score higher on measures of this sort of EI tend to be better leaders,[5,6] and to perform their jobs better than people who score lower,[7] especially when the jobs require a high degree of interpersonal interaction.[8]

That said, studies that focus on Goleman's model have shown that leaders with higher EI, as defined by that model, are often more effective and better at managing teams.[9,10,11] In addition, training the competencies that underlie Goleman's five EI domains has been linked to improved employee performance and productivity.[12]

Similarly, higher scores on a questionnaire that assesses Trait EI are correlated with better coping strategies, higher life satisfaction, better relationships, and better stress management.[13,14,15,16]

Finally, researchers report that measures based on Bar-On's mixed model of EI are related to mental health outcomes, stress management, team dynamics, and leadership outcomes.[17,18,19,20]

Although the models overlap to some extent, they tend to focus on different phenomena. For instance, the Four Branch model is about abilities per se, whereas the Trait Model is about something akin to a Big Five personality trait. I entered Table 7.1 into GPT-4 and asked it to identify the common threads that run through the models. Here's how it replied:

Me:

The attached table summarizes key aspects of different models of Emotional Intelligence. Based on what is in the table and what you can find elsewhere, could you please identify the common threads that run through the models. If you are quoting from other sources, please provide the citations. Thank you.

AI:

Based on the table and additional information from the field of Emotional Intelligence (EI), here are some common threads that run through the different models:

1. Emotion Recognition and Management: Across all models, there's a consistent emphasis on the abilities to accurately perceive and effectively manage emotions. This includes understanding one's own emotions and those of others, and applying this understanding to manage personal and interpersonal situations effectively.

2. Impact on Cognitive Processes: Several models highlight how emotions can influence cognitive processes. For example, the Four-Branch Model specifically discusses using emotions to enhance cognitive activities such as problem-solving and reasoning. This demonstrates a recognition that emotions can play a significant role in enhancing intellectual capabilities.

3. Social and Interpersonal Skills: A recurring element in these models is the importance of social and interpersonal skills. Goleman's model, for instance, incorporates social skills as a core domain, underlining the role of EI in effective leadership and management within organizational settings.

4. Overlap with Personality Traits: A common criticism across the models is their overlap with traditional personality traits. The Trait EI Model is particularly noted for blending aspects of personality traits such as extraversion and neuroticism into its structure, which may challenge the distinctiveness of EI as a unique construct.

5. Measurement Challenges: Concerns about the measurement of EI and potential biases in self-assessment are noted in several models. For example, both the Trait EI Model and Bar-On's Mixed Model utilize self-report questionnaires, which can introduce personal bias and affect the reliability and accuracy of EI assessments.

These commonalities suggest that while the conceptualization of EI may vary slightly from model to model, there is a strong consensus on the core components of emotion perception, management, and the utilitarian role of emotions in enhancing cognitive and social functioning. Additionally, the challenges related to measuring EI and distinguishing it from personality traits remain significant issues within the field.

The key takeaway is... "there is a strong consensus on the core components of emotion perception, management, and the utilitarian role of emotions in enhancing cognitive and social functioning."

The first time I ran this prompt, the results for each topic seemed to be written in different styles—which made me worry that GPT-4 might be sending me sections of its training materials. Hence, I asked for citations if it was quoting other sources. This simple addition to the prompt led to a very different result. But, applying a little critical thinking here, we must note that generative AIs usually provide different results even when the identical prompt is repeated, and thus it's difficult to know the causal relationship between the intervention and the results. Nevertheless, as noted previously, whenever there's reason to suspect that an AI is not accurate—or is doing something untoward—it's worth cycling through the CAL.

EI and Professional Success

It's not surprising that EI plays a key role in success at work, especially in jobs that require being able to interact effectively with others.[21] Indeed, employees with higher EI perform better at jobs that are high in "emotional labor," which are jobs that require frequent interactions with others (such as sales, nursing, teaching), require displaying certain emotions (usually positive), and require regulating emotions as part of the job.[22] Similarly, employees with high EI are more likely to receive promotions and ascend to higher levels within their organizations.[23]

The positive impact of EI on professional success may reflect different factors. For one, people who have higher EI tend to perform better academically.[24,25,26] And academic success is in turn associated with professional success. This relationship might indicate that a third variable underlies both effects, such as the possibility that people who have higher EI are more strongly motivated to succeed or are able to focus more effectively.

The fact that people high in EI are more successful professionally may also reflect enhanced interpersonal skills. In fact, people who have high EI are better at resolving conflicts.[27] This skill clearly draws on the ability to understand and manage emotions. EI also affects how well teams operate.[28,29] Not surprisingly, if the overall level of EI in the team is high, the team members are more likely to collaborate—and to perform better.

Some of the performance boost associated with high EI may also reflect how well people manage stress.[30,31] Indeed, people with higher EI tend to have more positive moods and positive self-esteem.[32] But the relation between EI and mental health may run deeper than that. For example, lower EI is associated with higher levels of depression and anxiety.[33] Of course, this is a correlation and we cannot know the direction of causality. The finding may indicate that EI can buffer against situations that lead to mental health challenges, or it may indicate that EI is difficult to acquire and cultivate if we are suffering from depression or anxiety—or both effects can be at work.

Most research on EI rests on correlational studies, in which researchers simply determine whether scores on EI questionnaires are associated with various other measures. As noted earlier, in such studies we don't know which variable is the cause and which is the effect—and it's possible that a third variable underlies any observed relationship. In contrast, training studies, in which some participants are trained on EI and compared to other similar people who are not trained, can allow us to identify EI as the active ingredient. And in fact numerous researchers report that they can train EI, in various contexts.[34,35,36,37,38] Moreover, research has documented that training EI can enhance work performance, conflict management, teamwork, employability, and job satisfaction.[39]

However, the story that has emerged from training studies is not simple. For example, Raquel Gilar-Corbi and colleagues[40] recruited senior managers of a private company to participate in a business-oriented EI training program. They first administered tests of different aspects of EI, and then had the participants engage in 30 hours of training, after which they administered the EI tests again. They compared their training group to a control group that did not receive training. They found that some aspects of EI improved after their training, but others did not. Specifically, they found that "intrapersonal EI, self-perception, general mood, self-expression, and stress management" improved in the training group but not the control group, and they found that "improvements in emotional understanding and emotion management" actually continued to improve one year after training had stopped. In contrast, interpersonal and adaptability skills did not change in either group.

Without question, EI is not a "single thing," but rather is a composite of different skills. Different training programs appear to exercise and improve different aspects of EI.

Using the CAL to Train EI

On the face of things, AIs seem unsuited for training EI because they cannot really "understand" all relevant aspects of many situations and contexts. However, AIs can familiarize people with the basics, and make them sensitive to the features of EI. Thus, I asked Claude 3 Opus to design a game to train key aspects of EI, as documented in the following. (A side note: For some reason, this AI initially wrote in another language, so I had to explicitly request English.)

Me:

Please conduct an interactive, real-time game to train a single user to have better emotional intelligence. Actually conduct the game with the user; don't just describe it. The game should have specific instructions, detailed scenarios, and clear ways to assess whether it is effective. Use principles from the attached table and other information you have to underlie the game. Please conduct this game in English. [ATTACH TABLE 7.1]

AI:

Alright, let's embark on an interactive game that will help you develop your emotional intelligence skills! We'll draw upon the key principles from the models discussed in the table, focusing on perceiving, understanding, and managing emotions in various scenarios.

Game: Emotion Navigator

Objective: Enhance your ability to recognize, interpret, and respond to emotions in yourself and others through real-life scenarios.

Instructions:

1. I will present you with a series of scenarios.
2. For each scenario, identify the emotions involved and suggest an emotionally intelligent response.

3. I will provide feedback and additional insights based on your answers.
4. Reflect on how you can apply these skills in your daily life.

Let's begin!

Scenario 1:

You're in a team meeting, and one of your colleagues expresses frustration about a project's progress. They appear visibly upset and direct their frustration towards the team.

Questions:

a) What emotions do you perceive in your colleague?
b) How would you respond to this situation in an emotionally intelligent manner?

Please provide your answers, and we'll discuss them together.

Me:

a) Anxiety, fear, and anger. b) I would note that we are all in this together, and need to help each other. Thus, the first step is to identify exactly what problem we need to solve. I would then ask the group for suggestions, and be sure to respect what everyone says.

AI:

Great observations and response! Let's break it down:

a) You correctly identified the key emotions:

- Frustration: Your colleague is expressing their dissatisfaction with the project's progress.
- Anger: They seem to be directing their negative emotions towards the team.
- Anxiety and fear: These emotions may be underlying their reaction, stemming from concerns about the project's outcome or their own performance.

b) Your response demonstrates several aspects of emotional intelligence:

- Empathy: You acknowledge that everyone is facing the same challenges, promoting a sense of unity and understanding.
- Problem-solving: By focusing on identifying the specific problem, you redirect the energy towards finding a solution rather than dwelling on negative emotions.
- Collaboration: Asking the group for suggestions encourages open communication and values diverse perspectives.
- Respect: Emphasizing the importance of respecting everyone's input creates a safe and inclusive environment for sharing ideas.

Reflection:

In this scenario, you showcased the ability to perceive and understand emotions in others, as well as the capacity to manage emotions in a constructive manner. By promoting empathy, collaboration, and problem-solving, you demonstrate key aspects of emotional intelligence that can contribute to a more positive and productive team dynamic.

Consider how you can apply this approach in other situations where emotions run high, such as conflicts with friends or family members. Remember that acknowledging emotions, focusing on solutions, and valuing others' perspectives are crucial skills in navigating emotionally charged situations.

Let's move on to the next scenario.

Scenario 2:

You receive an email from a client expressing dissatisfaction with a project you recently completed for them. They use strong language and question your competence in handling their account.

Questions:

a) What emotions might you experience upon reading this email?
b) How would you manage your own emotions and respond to the client in an emotionally intelligent way?

Me:

a) I would probably first get angry, then depressed, and then anxious. b) I would take a deep breath, and then go to an AI (like you), describe the situation, and ask for advice. I realize that your limited experience and abilities will prevent you from necessarily solving the problem, but I hope that what you say will spark ideas in me that will lead to some constructive response.

AI:

Your self-reflection and approach to managing your emotions in this scenario are commendable. Let's explore your response further:

a) Recognizing your own emotions is a critical aspect of emotional intelligence:

- Anger: It's natural to feel defensive or upset when your competence is questioned.
- Depression: Negative feedback can lead to feelings of sadness or disappointment in oneself.
- Anxiety: The uncertainty of how this situation may impact your professional reputation or client relationship can cause anxiety.

Acknowledging these emotions allows you to process them consciously rather than reacting impulsively.

b) Seeking advice from an AI is an innovative approach to gaining perspective and generating ideas. While AI may have limitations, it can offer valuable insights and prompt you to consider alternative viewpoints. Some suggestions an AI might provide:

- Take time to compose yourself before responding to the client. Avoid reacting while emotionally charged.
- Seek to understand the client's perspective. What specific concerns do they have, and how can you address them?
- Respond professionally and empathetically. Acknowledge their dissatisfaction, express your commitment to finding a solution, and propose steps to resolve the issue.

- Learn from the experience. Reflect on any areas for improvement in your work or communication with clients.

Remember, the goal is not to suppress your emotions but to manage them effectively. By taking a step back, seeking guidance, and approaching the situation with empathy and professionalism, you demonstrate emotional intelligence in handling a challenging client interaction.

Reflection:
In this scenario, you showcased self-awareness by recognizing your emotional response and the wisdom to seek external perspective. Engaging with an AI or other resources can provide valuable insights and help you approach the situation objectively.

Consider how you can incorporate this practice of self-reflection and seeking guidance in other emotionally charged situations, such as receiving critical feedback from a supervisor or navigating conflicts with colleagues. Remember that emotional intelligence involves not only managing your own emotions but also understanding and responding to others' emotions in a constructive manner.

You've done an excellent job engaging with these scenarios and demonstrating key aspects of emotional intelligence. Continue to practice these skills in your daily life, and remember that developing emotional intelligence is an ongoing journey.

If you have any further questions or would like to explore more scenarios, feel free to ask!

Me:

Thank you.

AI:

You're welcome! It's been a pleasure exploring these scenarios with you and discussing strategies to develop emotional intelligence.

Remember, enhancing your emotional intelligence is a continuous process that requires self-awareness, empathy, and effective emotion management. By consistently applying the principles we discussed, such as acknowledging your own emotions, seeking to understand others' perspectives, and responding in a constructive manner, you can strengthen your emotional intelligence skills over time.

Feel free to reach out if you ever want to discuss more scenarios, explore additional aspects of emotional intelligence, or seek guidance on navigating emotionally charged situations. I'm here to support you in your personal and professional growth.

Keep up the great work, and remember to be kind to yourself as you continue on this journey of self-improvement!

Please take another look at Scenario 1 in the above transcript, which specifies "You're in a team meeting, and one of your colleagues expresses frustration about a project's progress. They appear visibly upset and direct their frustration towards the team." I assumed that "frustration" wasn't going to be one of the key emotions, given that it was explicitly mentioned as part of the description of the situation. So, I avoided mentioning it in my response. But even so, the AI then responded to me "You correctly identified the key emotions: Frustration: Your colleague is expressing their dissatisfaction with the project's progress..." I did not say anything about frustration! This glitch underscores the need not to take an AI's responses at face value. We need to treat them as the basis for something we eventually create by revising and refining the result we obtain from the AI (see Chapter 2). Indeed, I treated each of its responses as an opportunity to gauge my emotional responses, as described in Chapter 5—and paid attention to those responses as the main way to make me more sensitive to the AI's suggestions.

Using the CAL to Boost Components of EI

It seems unlikely that current AIs can train EI beyond a basic level, but even that may sometimes be helpful. But rather than train EI in general, AIs might be more useful for training individual components of this complex ability.

For example, although recognizing emotions is only one small piece of EI, it is nevertheless critical. If we fail to recognize and interpret emotions correctly, we cannot reliably draw on other aspects of EI. And an AI can be trained to recognize facial expressions, which is one aspect of recognizing emotions. Indeed, several companies market AI products that interpret facial expressions, with an eye toward providing pointers to human users.[41] A good example of an existing product is offered by Affectiva, which has built software that automatically interprets facial expressions and facial movement in real time. This software can help people interpret how team members are feeling during meetings. This is an example of how an AI can relieve a burden from a human user. For instance, the AI could be prompted to flag when team members are looking anxious, which could in turn help the leader know when to slow down and clarify their communication—perhaps even pausing to ask a question of the audience, to identify the source of concern. Whatever its current limitations, software like this is a sign of what is to come.

Here's a second example of how AIs can help us with one component of EI: We can utilize an AI to provide feedback on how how we "read" another person's emotions during a social interaction, even if we must take its comments with ample amounts of salt. Such feedback should be used primarily to spark reflection, to help us look at things from a different perspective. In this case, we can ask AI to give us interpretations of another person's emotional reactions during a social interaction, and see what seems to ring true. Here's a prompt that I submitted to GPT-4o, which responded as below:

Please ask the user for an example of a social interaction that is puzzling them. Ask the user to describe it in detail. Wait until after the user has responded before continuing; do not continue until the user has finished responding. After this, ask the user direct questions in order to draw out additional details about the interaction. Your job is to try to find out what "really happened." Again, wait until the user has fully responded before continuing. After you have asked all relevant questions, please draw on your knowledge of "Emotional Intelligence" to provide two possible interpretations for this interaction.

AI:

Could you provide an example of a social interaction that is puzzling you? Please describe it in detail.

Me:

I was at a retirement party for a colleague and another colleague, who I didn't know very well, approached me. She was extremely tentative and nervous, and I couldn't tell why. I asked her where she lived and what she was doing now, and she answered very curtly and seemed irritated. I was making small talk in an effort to make her relax, but it seemed to backfire and I don't know why. Do you have any ideas?

AI:

Thank you for sharing that. To understand the situation better, can you provide some additional details?

1. Can you describe the setting of the retirement party? Was it formal or informal, and were there many people around you at the time of the interaction?
2. How did your colleague initially approach you? What was her body language like?
3. Can you recall the exact words or phrases she used when responding to your questions?
4. Did anything specific happen before or after this interaction that might have influenced her mood or behavior?
5. How did the interaction end? Did she walk away abruptly, or was there any further conversation?

These details will help me better understand what might have been happening in that interaction.

Me:

1. The setting during the conversation was informal. 2. She hovered around as I was talking to someone else, and then introduced herself—I had never actually talked to her before. 3. No, I cannot recall the exact words or phrases. 4. I had just given a speech thanking the retiring person for doing a great job. 5. The interaction just petered out. She was too nervous to say much and somebody else came over and joined us, at which point she drifted away.

AI:

Thank you for providing more details. Here are a couple of possible interpretations based on what you've described, using principles of emotional intelligence:

Interpretation 1: Social Anxiety or Introversion

The colleague who approached you might have social anxiety or be more introverted, which can make initiating conversations difficult and stressful. Her nervousness and tentative behavior could stem from discomfort in social settings, especially if she's not familiar with many people at the event. Your attempts at small talk, while well-intentioned, might have inadvertently added to her anxiety if she felt pressured to engage in conversation.

Interpretation 2: Misinterpretation of Intentions

She might have misinterpreted your questions about where she lived and what she was doing now as overly intrusive or personal, especially since you didn't know each other well. This can sometimes happen in social interactions where one person's friendly curiosity is perceived differently by the other person. Her curt responses and irritation could be a reaction to feeling like her personal boundaries were being crossed, even if that wasn't your intention.

In both cases, it seems like there was a disconnect between your intentions and her perceptions. Improving emotional intelligence in such scenarios involves being mindful of the other person's body language and emotional cues, and adjusting your approach accordingly. If someone seems particularly nervous or uncomfortable, it might help to keep the conversation light and non-personal until they seem more at ease.

Me:

I thought those questions were light, even if they were a bit personal. What should I have asked instead?

AI:

Given the context and her apparent nervousness, here are some alternative questions and conversation starters that might have made her feel more at ease:

1. General Observations about the Event:

   • "It's a great turnout for the retirement party, isn't it?"
   • "Have you been to one of these events before?"

2. Neutral Topics:

- "How do you know [retiring colleague]?"
- "What do you think of the venue?"

3. Shared Experiences:

- "Did you catch [a specific detail or moment from the event] earlier? It was quite something!"
- "What did you think of the speech/food/music?"

4. General Work-Related Topics:

- "How long have you been working here?"
- "What projects or teams are you currently involved with?"

These types of questions are less personal and focus more on shared experiences or general topics, which can help create a more comfortable atmosphere for someone who might be feeling nervous or uneasy.

Me:

That's helpful. Thank you.

AI:

You're welcome! I'm glad I could help. If you have any more questions or need further assistance with social interactions or anything else, feel free to ask.

This was not a profound interaction, but it did alert me to what to pay attention to in future interactions like this, and did give me some useful strategies I could employ. Thus, this exchange with the AI was helpful.

Flourishing with the CAL

EI can help us flourish in the Age of AI in several ways. For one, because it can help us navigate social situations successfully, it can foster a sense of autonomy and control. In addition, it clearly plays a role in having fulfilling relationships. Indeed, it is difficult to imagine having such relationships without EI. We cannot overstate how important having fulfilling relationships will be, going forward, if AIs do in fact reduce our workloads and give us more free time. If this comes to pass, we will spend more time interacting with others, and such interactions will probably play an increasingly central role in our lives.

We've alluded to the fact that one reason EI is so important is because it underlies much of effective human communication. We need to understand our own mental and emotional states when we communicate with others. Moreover, we need to take the perspectives of others in order to communicate effectively with them. In the following chapter we see how EI, critical thinking, creative thinking, managing cognitive and emotional constraints, and navigating personalities all come together when we communicate with one another.

Digging Deeper

The videos listed below provide accessible, more in-depth treatment of different aspects of EI. In addition, you might want to try using the search terms to locate new videos on the topic.

Emotional Intelligence by Daniel Goleman ▶ Animated Book Summary: https://www.youtube.com/watch?v=n6MRsGwyMuQ
   The Importance of Emotional Intelligence in the Age of AI: https://www.youtube.com/watch?v=pO6m44oFqvQ

- "Models of emotional intelligence"
- "Mayer and Salovey's Four-Branch Model of Emotional Intelligence"
- "Daniel Goleman's Emotional Intelligence Model"
- "Petrides and Furnham's Trait Model of Emotional Intelligence"
- "Bar-On's Mixed Model of Emotional Intelligence"
- "Emotional intelligence and job performance"
- "Emotional intelligence and leadership effectiveness"
- "Emotional intelligence and stress management"
- "Training emotional intelligence skills"
- "AI-based games for emotional intelligence training"
- "Challenges in training emotional intelligence with AI"

Emotional Intelligence: From Theory to Everyday Practice: https://www.youtube.com/watch?v=e8JMWtwdLQ4
   Emotional intelligence: What It Is and Why It Matters: https://www.youtube.com/watch?v=tN6mKvlxeBg

- "Emotions and cognitive processes"

- "Impact of emotions on decision making"
- "Emotional regulation strategies"
- "Emotional intelligence and interpersonal relationships"
- "AI-assisted emotional regulation techniques"
- "Mindfulness and emotional regulation"
- "Cognitive-behavioral therapy for emotional management"

References

1   Mayer, J. D., and Salovey, P. (1993). The intelligence of emotional intelligence. *Intelligence, 17*, 433–442. https://doi.org/10.1016/0160-2896(93)90010-3

2   Mayer, J. D., & Salovey, P. (1997). What is emotional intelligence? In P. Salovey & D. J. Sluyter (Eds.), *Emotional development and emotional intelligence: Educational implications*. Basic Books, pp. 3–31.

3   Mayer J. D., Salovey P., & Caruso D. R. (2004). Emotional intelligence: Theory, findings, and implications. *Psychological Inquiry, 15*, 197–215. https://doi.org/10.1207/s15327965pli1503_02

4   Lopes, P. N., Brackett, M. A., Nezlek, J. B., Schütz, A., Sellin, I., & Salovey, P. (2004). Emotional intelligence and social interaction. *Personality and Social Psychology Bulletin, 30*, 1018–1034. https://doi.org/10.1177/0146167204264762

5   Rosete, D., & Ciarrochi, J. (2005). Emotional intelligence and its relationship to workplace performance outcomes of leadership effectiveness. *Leadership & Organization Development Journal, 26*, 388–399. https://doi.org/10.1108/01437730510607871

6   Kerr, R., Garvin, J., Heaton, N., & Boyle, E. (2006). Emotional intelligence and leadership effectiveness. *Leadership & Organization Development Journal, 27*, 265–279. https://doi.org/10.1108/01437730610666028

7   O'Boyle Jr., E. H., Humphrey, R. H., Pollack, J. M., Hawver, T. H., & Story, P. A. (2011). The relation between emotional intelligence and job performance: A meta-analysis. *Journal of Organizational Behavior, 32*, 788–818. https://doi.org/10.1002/job.714

8   Joseph, D. L., & Newman, D. A. (2010). Emotional intelligence: An integrative meta-analysis and cascading model. *Journal of Applied Psychology, 95*, 54–78. https://doi.org/10.1037/a0017286

9   Goleman, D. (1996). Emotional intelligence: Why it can matter more than IQ. *Learning, 24*, 49–50.

10   Goleman, D., Boyatzis, R., & McKee, A. (2013). *Primal leadership: Realizing the power of emotional intelligence*. Harvard Business School Press.

11   Boyatzis, R. E., & Saatcioglu, A. (2008). A 20-year view of trying to develop emotional, social and cognitive intelligence competencies in graduate management education. *Journal of Management Development, 27*, 92–108. https://doi.org/10.1108/02621710810840785

12   Goleman, D. (1998). *Working with emotional intelligence*. Bantam Books.

13   Petrides, K. V., & Furnham, A. (2001). Trait emotional intelligence: Psychometric investigation with reference to established trait taxonomies. *European Journal of Personality, 15*, 425–448. https://doi.org/10.1002/per.416

14   Petrides, K. V., Pita, R., & Kokkinaki, F. (2007). The location of trait emotional intelligence in personality factor space. *British Journal of Psychology, 98 (Pt 2)*, 273–289. doi:10.1348/000712606X120618.

15   Gugliandolo, M. C., Costa, S., Cuzzocrea, F., Larcan, R., & Petrides, K. V. (2015). Trait emotional intelligence and behavioral problems among adolescents: A cross-informant design. *Personality and Individual Differences, 74*, 16–21. https://doi.org/10.1016/j.paid.2014.09.032

16   Mavroveli, S., Petrides, K. V., Rieffe, C., & Bakker, F. (2007). Trait emotional intelligence, psychological well-being and peer-rated social competence in adolescence. *British Journal of Developmental Psychology, 25*, 263–275. https://doi.org/10.1348/026151006X118577

17   Bar-On, R. (1997). *The emotional quotient inventory (EQ-i): A test of emotional intelligence.* Multi-Health Systems.

18   Bar-On, R. (2000). Emotional and social intelligence: Insights from the Emotional Quotient Inventory (EQ-i). In R. Bar-On & J. D. A. Parker (Eds.), *The handbook of emotional intelligence: Theory, development, assessment, and application at home, school, and in the workplace.* Jossey-Bass/Wiley, pp. 363–388.

19   Bar-On, R. (2006) The Bar-On model of emotional-social intelligence (ESI). *Psicothema, 18*, 13–25. https://www.psicothema.com/pdf/3271.pdf

20   Dawda, D., & Hart, S. D. (2000). Assessing emotional intelligence: Reliability and validity of the Bar-On Emotional Quotient Inventory (EQ-i) in university students. *Personality and Individual Differences, 28*, 797–812. https://doi.org/10.1016/S0191-8869(99)00139-7

21   Brasseur, S., Gregoire, J., Bourdu, R., & Mikolajczak, M. (2013). The Profile of Emotional Competence (PEC): Development and validation of a self-reported measure that fits dimensions of emotional competence theory. *PLoS ONE, 8*(5), e62635. https://doi.org/10.1371/journal.pone.0062635

22   Joseph, D. L., & Newman, D. A. (2010). Emotional intelligence: An integrative meta-analysis and cascading model. *Journal of Applied Psychology, 95*, 54–78. https://doi.org/10.1037/a0017286

23   Dries, N., & Pepermans, R. (2008). "Real" high-potential careers: An empirical study into the perspectives of organisations and high potentials. *Personnel Review, 37*, 85–108. https://doi.org/10.1108/00483480810839987

24   Costa, A., & Faria, L. (2015). The impact of emotional intelligence on academic achievement: A longitudinal study in Portuguese secondary school. *Learning and Individual Differences, 37*, 38–47. https://doi.org/10.1016/j.lindif.2014.11.011

25   Parker, J. D. A., Summerfeldt, L. J., Hogan, M. J., & Majeski, S. A. (2004). Emotional intelligence and academic success: Examining the transition from high school to university. *Personality and Individual Differences, 36*, 163–172. https://doi.org/10.1016/S0191-8869(03)00076-X

26   Sánchez-Álvarez, N., Berrios, M. P. B., & Extremera, N. (2020). A meta-analysis of the relationship between emotional intelligence and academic

performance in secondary education: A multi-stream comparison. *Frontiers in Psychology*, *11*, www.frontiersin.org/journals/psychology/articles/10.3389/fpsyg.2020.01517

27   Jordan, P. J., & Troth, A. C. (2004). Managing emotions during team problem solving: Emotional intelligence and conflict resolution. *Human Performance*, *17*, 195–218. https://doi.org/10.1207/s15327043hup1702_4.

28   Druskat, V. U., & Wolff, S. B. (2001). Building the emotional intelligence of groups. *Harvard Business Review*, *79*, 80–90.

29   Wolff, S. B., Pescosolido, A. T., & Druskat, V. U. (2002). Emotional intelligence as the basis of leadership emergence in self-managing teams. *The Leadership Quarterly*, *13*, 505–522. doi:10.1016/S1048-9843(02)00141-8.

30   Doğru, C. (2022). A meta-analysis of the relationships between emotional intelligence and employee outcomes. *Frontiers in Psychology*, *13*, www.frontiersin.org/journals/psychology/articles/10.3389/fpsyg.2022.611348

31   Mikolajczak, M., Roy, E., Luminet, O., Fillée, C., & de Timary, P. (2007). The moderating impact of emotional intelligence on free cortisol responses to stress. *Psychoneuroendocrinology*, *32*, 1000–1012. doi:10.1016/j.psyneuen.2007.07.009.

32   Schutte, N. S., Malouff, J. M., Simunek, M., McKenley, J., & Hollander, S. (2002). Characteristic emotional intelligence and emotional well-being. *Cognition and Emotion*, *16*, 769–785. doi:10.1080/02699930143000482.

33   Fernández-Berrocal, P., & Extremera, N. (2006). Emotional intelligence and emotional reactivity and recovery in laboratory context. *Psicothema*, *18(Suppl)*, 72–78.

34   Boyatzis, R. E., Goleman, D., & Rhee, K. S. (2000). Clustering competence in emotional intelligence: Insights from the Emotional Competence Inventory. In R. Bar-On & J. D. A. Parker (Eds.), *The handbook of emotional intelligence: Theory, development, assessment, and application at home, school, and in the workplace.* Jossey-Bass/Wiley, pp. 343–362.

35   Clarke, N. (2010). Developing emotional intelligence abilities through team-based learning. *Human Resource Development Quarterly*, *21*, 119–138. https://doi.org/10.1002/hrdq.20036

36   Durham, M. R. P., Smith, R., Cloonan, S., Hildebrand, L. L., Woods-Lubert, R., et al. (2023). Development and validation of an online emotional intelligence training program. *Frontiers in Psychology*, *14*, 1221817. doi:10.3389/fpsyg.2023.1221817.

37   Nelis, D., Quoidbach, J., Mikolajczak, M., & Hansenne, M. (2009). Increasing emotional intelligence: (How) is it possible? *Personality and Individual Differences*, *47*, 36–41. https://doi.org/10.1016/j.paid.2009.01.046

38   Nelis, D., Kotsou, I., Quoidbach, J., Hansenne, M., Weytens, F., Dupuis, P., & Mikolajczak, M. (2011). Increasing emotional competence improves

psychological and physical well-being, social relationships, and employ-ability. *Emotion, 11*, 354–366. https://doi.org/10.1037/a0021554

39   Kotsou, I., Mikolajczak, M., Heeren, A., Gregoire, J., & Leys, C. (2019). Improving emotional intelligence: A systematic review of existing work and future challenges. *Emotion Review, 11*(2), 151–165. https://doi.10.1177/1754073917735902

40   Gilar-Corbi, R., Pozo-Rico, T., Sanchez, B., & Castejon, J.-L. (2019). Can emotional intelligence be improved? A randomized experimental study of a business-oriented EI training program for senior managers. *PLoS ONE, 14*, e0224254. https://doi.org/10.1371/journal.pone.0224254

41   Green, F. M. (2023). How AI can help you develop emotional intelligence. *Forbes Newsletters*. www.forbes.com/sites/forbescoachescouncil/2023/03/24/how-ai-can-help-you-develop-emotional-intelligence/

Chapter 8

Enhancing Human Communication

Researchers at LinkedIn analyzed data from over 1 billion members to discover which skills were most in demand. The clear no. 1: communication skills. We need to know how to communicate effectively to different audiences, for different purposes, and in different media. Crafting clear and compelling communications is not easy in part because we must do this in open-ended situations in different contexts, to people who have different needs and backgrounds. Hence, LinkedIn noted "As organizations come to grasp the full extent of what AI *can* do, they're also coming to terms with all that it *can't* do—those tasks that require the uniquely human skills that all businesses need. That's the resounding takeaway from LinkedIn's latest global inventory of the most in-demand skills for professionals."[1] This conclusion jibes well with the central themes of this book, namely that: 1) we humans should double down on learning to deploy the skills needed to respond in open-ended situations that require taking context into account, and 2) we should use the Cognitive Amplifier Loop (CAL) to help us do so.

Our human interactions depend crucially on communication, and the techniques that underlie effective communication can be taught and learned. Emotional intelligence is part of this, but not the only part. To learn to be more effective communicators, we first must note the various types of communication humans engage in, and then consider the principles that govern communication and how to draw on them in different contexts.

Types of Communication

We humans communicate via a wide range of methods, each of which is appropriate for some purposes and not others. We can characterize the various methods of human communication based on five underlying dimensions. These dimensions help us understand the specific contexts and purposes for which each method is best suited, as well as their inherent strengths and limitations. These dimensions are as follows:

DOI: 10.4324/9781032686653-10

- *Formality.* Communication methods range from formal (e.g., formal teaching, webinars, presentations) to informal (e.g., conversations, text messages).
- *Synchronous vs. Asynchronous.* Communications can be synchronous, in real-time (e.g., conversations, theater productions, formal teaching) or asynchronous (e.g., email, blogs, literature).
- *Purpose and Content.* Communications can be intended to convey information (e.g., charts, graphs, formal teaching) or intended to persuade, entertain, or express emotion (e.g., songs, theater productions, works of art).
- *Mode of Delivery.* Communications differ in their modes of delivery. We can break this dimension down into three more granular dimensions: 1) Verbal communication (e.g., conversations, theater productions, podcasts) vs. non-verbal communication (e.g., body language, signage, and symbols). 2) Visual methods (e.g., videos, charts, graphs, diagrams, works of art) vs. auditory methods (e.g., music, podcasts). 3) Written methods (e.g., email, literature, blogs) vs. spoken methods (e.g., formal teaching, webinars). In some cases, two more or more of these modes may be combined (e.g., videos typically convey both visual and verbal information).
- *Complexity and Accessibility.* Finally, communications can be simple and easily accessible (e.g., text messages, signage, pop music) vs. complex, requiring specialized knowledge or skills (e.g., lectures on technical subjects, works of art, symphonic music).

Table 8.1 summarizes major methods of communication. Each of these communication methods relies on a unique combination of values on the five dimensions, which makes it appropriate for specific audiences and purposes. Choosing the appropriate method can be tricky, especially because Table 8.1 is only a top-level overview. The Table does not note many variants of each type—for example, written communications also include letters and magazines. Moreover, we can distinguish various versions of many of these categories, and they sometimes bleed into each other to create hybrids.

We would need to put in an enormous amount of work to master all of the ways that we humans can communicate, all of their potential purposes, and their strengths and limitations. Once again, we can use the CAL to help relieve the burden. If we aren't sure which communication method is most appropriate, we can try using the following prompt—which can build on the methods in Table 8.1 and produce new ones that fit our needs. In this example, I used Claude 3 Opus:

Table 8.1 Major Methods of Human Communication.

Type	Method	Summary	Purposes	Strengths	Limitations
Written Communication	Email	Written language delivered electronically.	Documentation, wide reach, asynchronous communication.	Records interactions, delivered quickly, accessible on many types of devices.	Lacks immediacy, can be impersonal, information overload.
	Text Messages	Short written messages exchanged electronically.	Quick and convenient.	Informality, immediacy, accessible on many devices.	Misinterpretation, superficial, overdependence.
	Literature	Written works, such as novels and poetry, which use language to communicate ideas and stories.	Knowledge dissemination, entertainment, cultural preservation.	Conveys content by "showing not telling."	Time-consuming to produce and consume, requires literacy.
	Blogs	Online written content that shares information or personal opinions.	Personal expression, information sharing.	Wide reach, accessible via many devices, cheap and easy to produce.	Requires writing skills, information overload, credibility issues.
Visual Communication	Charts and Graphs	Graphical representations of data.	Concise overview of patterns in data.	Effective summary of information, visual impact.	Can be misinterpreted, distorted, oversimplified, requires graphical literacy.

	Diagrams	Visual representations of relationships among elements.	Clear communication of processes, instructional aid.	Visual clarity.	May require technical understanding, can be oversimplified or too complex.
	Signage and Symbols	Visual symbols or signs that convey specific information.	Highly structured directions.	Efficient for conveying simple messages, useful in diverse settings.	May not be understood universally without context, oversimplification of complex messages.
Oral/Verbal Communication	Conversation	Direct verbal interaction between people.	Personal interaction, relationship building, conveying recent information.	Ease of producing immediate feedback.	Can be misinterpreted, requires both parties to be present at the same time, can be inefficient, transient.
	Presentations	Spoken communication, often supported by visual aids to inform or persuade an audience	Educational, persuasive.	Highly structured, can reach large audiences.	Requires public speaking skills, often minimally interactive, preparation intensive.
	Podcasts	Spoken word content made available digitally.	Personal expression, information sharing, entertainment.	Wide reach, accessible via many devices, cheap and easy to produce.	Limited to audio, requires engaging content to maintain interest, no visual aids.

(Continued)

Table 8.1 (Continued)

Type	Method	Summary	Purposes	Strengths	Limitations
	Dramas	Scripted spoken performances by actors.	Entertainment, cultural and moral reflection, emotional engagement.	Conveys content by "showing not telling."	Limited to scripted content, expensive to produce and present, may require audience sophistication.
	Webinars and Online Conferences	Online events that involve verbal presentations and discussions.	Educational, interactive.	Wide reach without geographical constraints.	Dependent on technology, minimal personal interaction, potential technical issues.
	Formal Teaching	Educational sessions where information is conveyed through spoken language.	In-depth knowledge sharing, structured learning.	Scaffolded instruction, mentorship, guidance, motivational support.	Can be rigid, limited to curriculum, may not cater to appropriate levels of learners.
Non-Verbal Communication	Body Language	Conveys feelings and intentions through physical behavior.	Subtle expression of emotions and attitudes, non-verbal cues.	Conveys emotional information that is difficult to put into words.	Can be misinterpreted, culturally dependent, often subconscious.
	Works of Art (Painting, Sculpture, Music, Dance)	Conveys meaning through visual and/or auditory experiences.	Emotional expression, indirect content.	Cultural and societal impact, aesthetic enjoyment.	Subjective interpretation, accessibility requires artistic literacy.

Me:

Begin by asking the user one question: Ask them to describe what they want to communicate. Wait for them to respond. After they respond, then ask follow-up questions, one at a time, based on the attached chart. After each question, wait for the user to respond before asking the next question. The answers to your questions should lead you to converge on advising a particular type of communication as most appropriate. Please suggest to the user that they use that modality, and explain why it is most appropriate. [ATTACH TABLE 8.1]

AI:

To help determine the most effective communication method for your needs, please describe what you want to communicate.

Me:

Sure. I want to create a message that will discourage people from littering.

AI:

Thank you for that information. To help narrow down the best communication method, could you please provide some more details about your target audience? For example, is this message intended for a specific age group, location, or demographic?

Me:

I'm particularly interested in people between the ages of 14 and 21, especially when they are in groups and likely to be distracted.

AI:

Given your target audience of 14 to 21-year-olds who are likely to be in groups and distracted, visual communication methods might be most effective. Do you have a preference for static visuals (like signage or posters) or more dynamic visuals (like short videos or animations)?

Me:

More dynamic would probably be more likely to get attention and be engaging.

AI:

Considering your target audience and preference for dynamic visuals, creating a short video or animation could be an effective way to communicate your anti-littering message. Videos can combine visual and auditory elements to capture attention and convey information in an engaging way.

Do you have any specific ideas for the content or style of the video that you think would resonate with your target audience?

Me:

I have only a minimal budget, so it needs to be simple. Perhaps using humor would be an effective way to convey the message.

AI:

Given your minimal budget and the desire to use humor, I would suggest creating a short, simple animated video or a series of brief, funny video skits. Here's why this approach could be effective:

1. Humor can make the message more memorable and engaging for your target audience.
2. Short, simple animations or skits can be produced on a low budget using basic tools or free online resources.
3. Videos can be easily shared on social media platforms popular with your target demographic, increasing the reach of your message.
4. The combination of visual and auditory elements can help capture attention, even when your audience is distracted.

When creating the content, consider using relatable scenarios, exaggerated consequences, or unexpected twists to drive home the anti-littering message in a humorous way. Keep the videos short (30–60 seconds) to maintain engagement and encourage sharing.

Does this seem like a good approach for your anti-littering communication? If so, I can provide some more specific ideas for the video content.

Me:

Yes, it seems fine. Do you think you could help me produce a short, simple animated video?

AI:

As an AI language model, I don't have the capability to directly produce animated videos. However, I can provide you with guidance and resources to help you create a simple animated video on a minimal budget. Here are some steps you can follow:

1. Scriptwriting: Develop a short script for your video that effectively conveys your anti-littering message through humor. Keep it concise and engaging.
2. Storyboarding: Create a simple storyboard to visualize the flow of your video and plan out the scenes, characters, and animations needed.
3. Choose a tool: Select a user-friendly animation tool that fits your budget and skill level. Some options include:

   - PowToon: A web-based tool with pre-made templates and characters for creating animated videos.
   - Vyond (formerly GoAnimate): Another web-based platform with a library of templates, characters, and props.
   - Animaker: A DIY animation software with a free version that allows you to create short videos.
   - Blender: A free, open-source 3D animation software that has a steeper learning curve but offers more flexibility.

4. Create assets: Design or select characters, backgrounds, and props for your animation using the chosen tool or free online resources like Freepik or Flaticon.

5. Animate: Use the tool to animate your storyboard, bringing your characters and scenes to life.
6. Add sound: Record or find free sound effects and music to enhance your video's impact. Ensure that you have the necessary permissions or licenses for any audio you use.
7. Export and share: Export your final video in a web-friendly format and share it on social media platforms and other channels relevant to your target audience.

Remember, the key is to keep the animation simple, short, and humorous to effectively engage your audience and convey your message. If you need any further assistance or have questions during the process, feel free to ask!

Me:

Thank you.

AI:

You're welcome! I'm glad I could help guide you through the process of creating a simple animated video for your anti-littering campaign. Remember to keep your message clear, concise, and engaging, and don't hesitate to reach out if you need any further assistance.

Once you've completed your video, consider promoting it through various channels such as:

1. Social media platforms (e.g., Instagram, TikTok, Facebook)
2. Schools and educational institutions
3. Local community organizations and events
4. Environmental groups and campaigns

By spreading your message through multiple channels, you can maximize your reach and impact on your target audience. Good luck with your project, and remember, every effort to create a cleaner, more sustainable environment counts!

As is evident, AIs can in fact help us devise interesting and potentially effective ways to communicate a specific message. Notice how creative it was; it did not simply spit back entries from Table 8.1, but instead built on them. Moreover, we can easily follow up and ask the AI to help us create memes, scripts, catchy hashtags, and even (sometimes) passable jokes.

Principles of Clear and Compelling Communications

All types of human communication must respect the fact that humans are the ultimate consumers of messages. This means that communications must respect the cognitive limitations of humans, not providing too much information at once and thereby overloading working memory, nor requiring multitasking and thereby exceeding our attentional capacities (see Chapter 5).

Researchers have documented a number of principles that can guide us to create clear, compelling, and memorable communications. We focus here on communications intended to convey information, which are particularly important when we need to work with other people. Many of the principles that guide how we communicate when we work with others are captured by the *Seven Cs of Communication*.[2] The Seven Cs are: Clear, Concise, Concrete, Correct, Coherent, Complete, and Courteous. We can use these guidelines as a checklist to ensure that communications are effective. This sort of check is almost always useful, particularly in professional and business contexts. Consider each principle in turn:

- *Clear.* Information-conveying communications should be as straightforward and cogent as possible. To achieve this, they should typically rely on precise, concrete, and unambiguous words or symbols.
- *Concise.* Einstein supposedly said "Make things as simple as possible, but no simpler." This definitely applies to communications. We should use as few words or symbols as necessary and avoid unnecessary redundancy. This practice not only respects the recipient's time, it also helps to maintain attention and prevent information overload.
- *Concrete.* Information-conveying communications should be specific and detailed, using facts, figures, and real-life examples to clarify the point.
- *Correct.* Messages should be factually accurate, and not include grammatical, formatting, or spelling errors. This practice is especially important in professional settings.
- *Coherent.* Communications should be logical and consistent. All parts of the communication should be connected and relevant to the topic.
- *Complete.* We should provide all the necessary information and not include irrelevant information.
- *Courteous.* We should be respectful, appropriate, and considerate. This requires not being aggressive or using offensive language or symbols.

Respecting Context

The Seven Cs checklist is useful, but the way we apply it depends on the precise context of the communication. For example, the philosopher H. P. Grice developed a set of four "Maxims" to describe how people should communicate in everyday conversations.[3] Some of these Maxims overlap with some of the Seven Cs, but may help to focus them further. Grice's four Maxims are as follows:

Maxim of Quantity. We should provide as much information as needed, but no more. This maxim emphasizes the importance of addressing a question or issue without overloading the recipient with overly detailed information. For example, if someone asks us what time it is, we should give them the hour and minutes, but not the seconds—which is overly detailed and irrelevant for most purposes.

Maxim of Quality. We should attempt to provide accurate information and avoid providing information that is false or not supported by evidence. For example, if we don't know the answer to a question, it's better to admit ignorance than to make up an answer. AIs are sometimes guilty of violating this maxim, given their penchant for hallucinating.

Maxim of Relevance. We should only provide relevant information. This maxim indicates that contributions to a conversation should be pertinent to the topic being considered. For example, if we are discussing climate change, we would violate this maxim if we suddenly started talking about behavioral issues at our children's school.

Maxim of Manner. We should avoid being ambiguous, vague, or obscure. This maxim is about being as clear and straightforward as possible, which requires us to organize our thoughts so that they are coherent. For example, if we give directions, we should provide them step-by-step in a way that is easy to follow.

If we violate these Maxims, the recipient may be confused or may look for hidden meanings or intentions, using the context to try to infer what's going on. For example, if we violate the Maxim of Relevance by abruptly switching the topic, the recipient might suspect that we are bored or nervous about the direction the conversation is taking.

Although Grice's Maxims have intuitive appeal and have been supported by some empirical research,[4] critics note the need to revise them in various ways (e.g.,[5,6]). For example, indirect and non-literal forms of speech, such as sarcasm or metaphor, may violate these Maxims but the recipient still knows how to grasp the point. In spite of these critiques, Grice's Maxims remain a fundamental framework for understanding the principles underlying straightforward communication in social interactions.

Now let's consider another factor that depends on the context of a social interaction: politeness. Being polite is a social lubricant, it is one of the mechanisms that allows us humans to get along with each other. But more than that, it is also a communication lubricant. Being polite makes people more receptive to what we want to convey. Indeed, there's even evidence that AIs respond better when the prompts are polite, and this is true in multiple languages.[7]

Penelope Brown and Stephen Levinson developed a detailed theory of how to incorporate politeness into communications.[8,9] Their theory is rooted in the concept of "face," which is the public image that each of us wants to project. The theory gives rise to communication strategies that maintain and respect the recipient's face. These strategies are designed to minimize potential conflicts and facilitate smoother social interactions.

The theory proposes that everyone has two basic types of face needs or desires. "Positive Face" implies that everyone wants to be appreciated, approved of, and valued by others. Communications that support a recipient's positive face include compliments, approval, and acknowledgments of the other person's needs and goals. "Negative Face" implies that everyone wants autonomy and wants their personal space respected. Communications that support these needs recognize and support the recipient's right to act freely and make their own decisions.

To be polite, communications should avoid threatening either the speaker's or the listener's positive or negative face. Brown and Levinson identify various "Face-Threatening Acts," such as making requests that require others to expend effort, giving orders, criticizing, or even offering compliments or accepting thanks when they impose on the autonomy of the other person. They propose specific strategies to mitigate or avoid these face threats. One is to show friendliness, interest, or invoke a sense of belonging, often by making appropriate compliments or expressing concern. In addition, being indirect, formal, or apologetic acknowledges that we know that we are imposing on others.

Brown and Levinson note that politeness is more or less important in different contexts. For example, the greater the social distance between the speaker and the recipient, the more important it is to rely on politeness strategies. Related to this, we should use politeness strategies when we communicate with someone who has more power than we do. And we should use such strategies if we are requesting something that requires substantial effort.

One of the main criticisms of this theory is that it reflects Western norms and assumptions, and may not apply to other cultures.[10,11,12] The theory has also been criticized because it focuses on transactional communications, and doesn't address more spontaneous interactions or interactions with people who know each other well and have a lot of history together. Again, context is crucial.

Getting Through

All of what we have discussed so far in this chapter can also help us to persuade others. However, in this case we need to draw on more than general communication skills, even when combined with critical thinking and creative problem-solving skills—we also must rely on emotional intelligence. For example, one crucial type of persuasion during the Age of AI will be convincing people not to fall prey to emotionally compelling disinformation and misinformation. It's clear that simply providing evidence of distortions or fabrications isn't enough to sway people's opinions or beliefs.[13] It's not enough to tell or even show people the error of their ways.

Instead, to dissuade people from believing emotionally compelling disinformation and misinformation, we need to rely on empathy, Theory of Mind, and the sort of human touch that stems from emotional intelligence. The results of studies suggest that to debunk such fabrications we must listen carefully and identify the underlying concerns that makes someone vulnerable to accepting false information. We then should acknowledge the grains of truth that make such information compelling.[14] Moreover, we need to engage others so that they not only feel heard, but also feel that they have played a role in interpreting the claim in question. Researchers have provided evidence that people are most likely to be convinced when they actively participate in constructing a narrative.[15,16] They need to feel emotionally engaged.

A Communications Facilitator and Editor

We can use AIs to help us write all manner of communications, ranging from emails, to marketing materials, to song lyrics, to bedtime stories. We can apply the communication guidelines we just reviewed to all of these modalities, to one degree or another. However, even the Seven C's and Grice's Maxims are a lot to keep in mind easily, which again invites using an AI as a cognitive amplifier.

Targeting a Specific Audience

We don't provide the same information, or present it in the same way, to all audiences. Rather, we need to adjust our communications to fit the context. For example, consider the following scenario:

A couple has just had a spat over a financial issue. The husband spent a lot of money on a set of golf clubs, which his spouse found problematic given their budget. Moreover, the husband had just refused to allow the spouse to purchase a gorgeous painting because it was too expensive. Worse still, the husband tried to conceal the price of the clubs, which the spouse stumbled upon when seeing a receipt that was carelessly left on a night table. The ensuing argument was acrimonious and full of angry finger-pointing. The spouse rushed out of the house, loudly slamming the front door.

Three people heard muffled shouting, and witnessed the spouse leaving in a huff. The husband comes out of the house, and over the course of the next hour three people ask him, individually, what happened: a neighbor, the neighbor's young daughter, and a passing law enforcement officer.

If you were the husband, how would you go about explaining this incident to each of them?

I submitted the following prompt to GPT-4.

I would like you to play the role of a husband who has to explain an incident to three different people. In all cases, please use the Seven C's of communication and please respect Grice's Maximums. Part 1 describes the situation. Part 2 describes the people to whom you should explain what happened. Please adapt your explanation so that it is appropriate for each of the three people noted in Part 2.

Part 1, the incident: [PASTE IN ABOVE SCENARIO]

Part 2, the people who want to know what happened: a neighbor, the neighbor's young daughter, and a passing law enforcement officer. You are playing the role of the husband, who has to explain the situation.

Here's how GPT-4 responded:

Explanation to a Neighbor
"Hey, I just wanted to clear the air about the noise earlier. My spouse and I had a disagreement about some financial decisions. I made a purchase that wasn't planned in our budget, and it understandably

upset her, especially since I had previously objected to a purchase she wanted to make. Things got a bit heated, and I'm sorry if it caused any disturbance."

Explanation to the Neighbor's Young Daughter

"Hi there, you might have heard some loud noises from our house today. Sometimes adults have disagreements about things like buying stuff—it's like when you really want a toy but have to wait. We just had a grown-up version of that. Everything is okay, and we're just trying to sort things out."

Explanation to a Passing Law Enforcement Officer

"Officer, thank you for checking in. We had a domestic dispute over some financial issues. I purchased something quite expensive without discussing it properly with my spouse, and it led to a significant argument. My spouse left to cool down after our disagreement escalated. We're handling it internally, and I assure you we're working to resolve the matter peacefully."

The AI did modulate the message more-or-less appropriately, and did draw on the Seven C's and Grice's Maxims—even though I didn't upload summaries of this material. In many situations, we don't need to upload material for guidance, but rather can rely on the training data that went into creating the AI. Such training data is probably sufficient if the material is very specific, concrete, and presented in a consistent way in numerous sources. Both the Seven C's and Grice's Maxims fit this bill. To check that the training data were accurate, I simply asked GPT-4 to give me quick summaries of the Seven C's and Grice's Maxims—which it did well. Thus, I was comfortable writing the above prompt, where I only referred to the key considerations, as opposed to uploading the corresponding tables.

However, as usual, we need to edit these drafts. For example, depending on the age of the young daughter, the reference to a "toy" might not be appropriate. In addition, we should edit for politeness, which can be subtle and nuanced.

In general, an AI can help us draft many kinds of messages for different audiences or recipients. For example, it would couch things differently for the CEO vs. a staff assistant. We can also utilize the CAL to help us draft personal communications, such as letters to an insurance company, doctor, family, and so on.

Table 8.2 Meyer's Eight Dimensions of Cultural Variations.

Dimension	Description	Example
Communicating	High-Context vs. Low-Context: High-context communication relies on implicit context and shared knowledge, often vague and indirect. Low-context communication is explicit and direct.	High-Context: Japan, Arab countries; Low-Context: United States, Germany.
Evaluating	Direct Negative Feedback vs. Indirect Criticism: Direct feedback is explicit and frank, whereas indirect criticism is subtle and often embeds negative feedback in a positive surrounding context.	Direct: United States; Indirect: Japan.
Persuading	Principles-First vs. Applications-First: Principles-first cultures use deductive reasoning, starting with theory. Applications-first cultures use inductive reasoning, starting with specifics. However, Asian cultures may not fit into this dimension because they approach situations holistically, not in terms of relations among components.	Principles-First: France, Italy; Applications-First: United States, Anglo-Saxon countries.
Leading	Egalitarian vs. Hierarchical: Egalitarian cultures have flat structures where everyone contributes, whereas hierarchical cultures have clear chains of command with top-down decisions.	Egalitarian: Scandinavian countries; Hierarchical: China, Japan.
Deciding	Consensual vs. Top-Down: Consensual decision making involves stakeholders and seeks agreement, whereas top-down decision making is leader-driven without much consultation.	Consensual: Japan, Netherlands; Top-Down: China, Nigeria.
Trusting	Task-Based vs. Relationship-Based: Task-based trust is earned through performance and results, whereas relationship-based trust is built through personal relationships and shared experiences.	Task-Based: United States, Germany; Relationship-Based: China, Brazil.
Disagreeing	Confrontation vs. Avoidance: Confrontational cultures openly debate and confront issues, whereas avoidance cultures seek harmony and address disagreements indirectly.	Confrontation: Israel, Germany; Avoidance: Japan, Sweden.
Scheduling	Linear-Time vs. Flexible-Time: Linear-time cultures adhere to strict schedules and deadlines, whereas flexible-time cultures value relationships over schedules and view time as fluid.	Linear-Time: Switzerland, United States; Flexible-Time: India, Saudi Arabia.

Communicating Across Cultures

It is no mean feat to characterize not only how culture affects communication, but also how communication in another culture differs from our own. To address these challenges, Erin Meyer published a groundbreaking book entitled *The Culture Map: Breaking Through the Invisible Boundaries of Global Business*.[17] Meyer defines eight dimensions that characterize how people in different cultures think and communicate, each of which grew out of prior theory and research. The dimensions are Communicating,[18] Evaluating,[19] Persuading,[20,21] Leading,[22] Deciding,[23] Trusting,[24] Disagreeing,[25,26] and Scheduling.[27] These dimensions are summarized in Table 8.2. We can use these dimensions to create a "map," which indicates specific ways that two cultures are similar and different. The CAL can help us respect such differences.

But first, a word of caution: Like all generalizations, there are exceptions within any given culture. The general tendencies noted in the dimensions not only vary within cultures but also can vary within individuals who are in different contexts. For example, an Indian working with a US company may treat time very differently than they do when they work with other Indians. Thus, an AI can only get us so far in respecting cultural differences.

That said, specific tendencies and predilections do generally characterize different cultures, and an AI can respect those characteristics to help us communicate to an audience in a different culture. We can engage an AI in this way by uploading Table 8.2 and asking the AI to draw on it when drafting a memo or document. For example, consider the following:

Me:

Please draft a memo that requests Department Heads to send in their annual report by 1 June. Explain to them that if we miss this deadline, we may not be eligible for important government funding. Thus, urge them to please turn in a complete, carefully drafted, report by the deadline. Please draft two versions of this memo, both of which are sent from Headquarters which is based in the United States. One memo is to be sent to the Department Head in Japan and one is to be sent to the Department Head in Germany. To adjust the memos appropriately for the different cultures, please draw on the uploaded table. Thank you. [ATTACH TABLE 8.2]

AI:

Here are two tailored memos for the Department Heads in Japan and Germany, based on the cultural dimensions provided in the uploaded document:

Memo for Department Head in Japan (High-context, Indirect, Consensual, Avoidance)

From: Headquarters, United States
Date: [Current Date]
To: [Department Head's Name], [Department Name]
Location: Japan

Subject: Gentle Reminder for the Timely Submission of Annual Report

Dear [Department Head's Name],

As we approach the conclusion of another productive year, we kindly remind you of the importance of submitting your department's annual report by the upcoming deadline of 1 June.

Your meticulous attention to detail and comprehensive review in preparing this report are crucial.

It is essential for us to adhere to this schedule as it significantly impacts our eligibility for vital government funding. We trust that with your cooperation and diligent efforts, we will successfully meet this deadline, ensuring the continuity and success of our initiatives.

We appreciate your understanding and collaboration. Should you need any assistance or clarification, please do not hesitate to reach out.

Warm regards,

[Your Name]
[Your Position]
Headquarters, United States

Memo for Department Head in Germany (Low-context, Direct, Confrontation, Task-based)

From: Headquarters, United States
Date: [Current Date]
To: [Department Head's Name], [Department Name]
Location: Germany

Subject: Immediate Action Required: Submission of Annual Report by 1 June

Dear [Department Head's Name],

This memo serves as a direct reminder that the deadline for submitting your department's annual report is approaching on 1 June. We urge you to prioritize the completion of this report as missing this deadline could jeopardize our eligibility for essential government funding.

It is imperative that your report be detailed, accurate, and submitted on time. Please ensure that all necessary data and analysis are comprehensive and reflective of our high standards.

We count on your usual efficiency and professionalism to meet this critical deadline. Please contact headquarters immediately if you foresee any challenges in meeting this deadline.

Thank you for your prompt attention to this matter.

Best regards,
[Your Name]
[Your Position]
Headquarters, United States

These memos are adjusted to respect and align with the cultural norms and communication styles of Japan and Germany, aiming for effectiveness in message delivery and compliance with organizational deadlines.

These drafts are a bit clunky, and definitely need a human touch to refine them—but the AI clearly can help us to respect cultural differences when we draft documents that are targeted to members of different cultures.

A Personal Editor

We can also utilize AIs to edit what we write. For example, we can submit a simple prompt like this:

> Attached please find a book chapter I've written. Could you please flag any typos and any situations where the language is unclear? Please also let me know whether I've respected the Seven C's of communication and Grice's Maxims. In addition, please indicate whether the organization works well, and if it does not please make suggestions about how to improve it.

When I submitted this prompt along with an early draft of the present chapter, GPT-4 offered valuable advice, not only about the details of sentence structure but about use of subheadings, examples, and other pointers intended to help with the flow. I did not agree with all of them, but found them reasonable and thought-provoking.

We can employ an AI to edit all manner of written communications, ranging from memos, to letters, to essays, to screenplays. The AI will do well in highly structured situations where the context is clear. However, to the extent that the situation is open-ended, and effects of context are complex or subtle, we need to provide specific guidance in addition to the general requests in the above prompt. For example, we would not want the AI to edit a screenplay for a comedy the same way we would want it to edit one for a tragedy. In such situations, we need to provide background information about the material, and often should provide examples of appropriate edits—as well as examples of what not to do (e.g., removing humor from a comedy!).

We can use similar prompts to edit illustrations. As we see in Table 8.1, writing is not the only way we communicate. And in fact, GPT-4o, MidJourney, Stable Diffusion, and various other generative AIs can help us produce many types of illustrations and graphics. Indeed, various software systems—such as Invideo.io, Kuaishou's Kling, Narakeet, and OpenAI's Sora—allow users to create high-quality videos based on verbal prompts. One virtue of using these programs, as opposed to stock photos or videos (e.g., such as those found on Pixabay), is that they are unique. Moreover, when we use an AI we can design videos to precisely illustrate a particular point. Although creating videos with an AI often requires many iterations of the CAL, it's remarkable what these models can produce if we are patient and persistent.

Flourishing with the CAL

Becoming a better communicator is key to many aspects of flourishing, both now and in the coming years. Clearly, being a better communicator enhances our feelings of autonomy and control, particularly in social situations. And these abilities are critical for having fulfilling relationships and succeeding in most jobs. But more than that, learning to communicate well is essential for personal growth and developing our abilities and talents. Moreover, these skills will help us develop appropriate life goals that give us a sense of purpose or meaning in life.

We have focused here primarily on situations where we need to communicate information in one go, and not on situations where there is a dynamic cycle of communication. Nevertheless, the same principles apply even in these sorts of situations. Indeed, as we see in the following chapter, these same principles apply to communications between leaders and followers and, in general, among people who are collaborating to reach a common goal.

Digging Deeper

The following videos provide a more in-depth treatment of different aspects of the material discussed in this chapter. As usual, the search terms can help to locate new videos on the topic.

Think Fast, Talk Smart: Communication Techniques: https://www.youtube.com/watch?v=HAnw168huqA
   Communication Skills—How to Improve Communication Skills—7 Unique Tips! https://www.youtube.com/watch?v=mPRUNGGORDo

- "Most in-demand skills for professionals"
- "Importance of communication skills in the workplace"
- "Types of human communication methods"
- "Formal vs. informal communication"
- "Synchronous vs. asynchronous communication"
- "Purposes and content of communication"
- "Modes of communication delivery"
- "Verbal vs. non-verbal communication"
- "Visual vs. auditory communication methods"
- "Written vs. spoken communication"
- "Complexity and accessibility of communication methods"
- "Choosing the appropriate communication method for the context"

The 7 Cs of Communication: https://www.youtube.com/watch?v=sYBw9-8eCuM

Grice's Maxims: https://www.youtube.com/watch?v=HcR9KYLuIGA

- "Principles of effective communication"
- "Seven Cs of Communication"
- "Grice's Maxims of conversation"
- "Maxim of Quantity, Quality, Relevance, and Manner"
- "Brown and Levinson's Politeness Theory"
- "Face-threatening acts in communication"
- "Politeness strategies in different cultures"
- "Context-dependent communication strategies"
- "Cognitive limitations and effective communication"

The Culture Map: The Future of Management: https://www.youtube.com/watch?v=qf1ZI-O_9tU

Cross cultural communication | Pellegrino Riccardi | TEDxBergen: https://www.youtube.com/watch?v=YMyofREc5Jk

- "Erin Meyer's Culture Map"
- "High-context vs. low-context communication cultures"
- "Direct negative feedback vs. indirect criticism cultures"
- "Principles-first vs. applications-first cultures"
- "Egalitarian vs. hierarchical cultures"
- "Consensual vs. top-down decision-making cultures"
- "Task-based vs. relationship-based trust cultures"
- "Confrontational vs. conflict-avoidant cultures"
- "Linear-time vs. flexible-time cultures"
- "Navigating cultural differences in communication"
- "Adapting communication styles across cultures"
- "Challenges and strategies for cross-cultural communication"

References

1   Brodnitz, D. (2024, February 8). The most in-demand skills for 2024. *LinkedIn Talent Blog*. www.linkedin.com/business/talent/blog/talent-strategy/linkedin-most-in-demand-hard-and-soft-skills

2   Baird, J., & Stull, J. (1992). *The seven C's of communication*. Prentice Hall.

3   Grice, H. P. (1975). Logic and conversation. In D. Davidson & G. Harman (Eds.), *The logic of grammar*. Dickenson, pp. 64–75.

4   Davies, C. & Katsos, N. (2013). Are speakers and listeners "only moderately Gricean"? An empirical response to Engelhardt et al. (2006). *Journal of Pragmatics, 49*, 78–106.

5   Hossain, M. M. (2021). The application of Grice Maxims in conversation: A pragmatic study. *Journal of English Language Teaching and Applied Linguistics, 3*, 32–40. doi:10.32996/jeltal.2021.3.10.4.

6   Kleinke, S. (2010). Speaker activity and Grice's maxims of conversation at the interface of pragmatics and cognitive linguistics. *Journal of Pragmatics, 42*, 3345–3366. https://doi.org/10.1016/j.pragma.2010.05.008

7   Yin, Z., Wang, H., Horio, K., Kawahara, D., & Sekine, S. (2024). *Should we respect LLMs? A cross-lingual study on the influence of prompt politeness on LLM performance.* Cornell University. *arXiv*:2402.14531v1.

8   Brown, P., & Levinson, S. C. (1978). Universals in language usage: Politeness phenomena. In E. N. Goody (Ed.), *Question and politeness: Strategies in social interaction.* Cambridge University Press, pp. 56–310.

9   Brown, P. & Levinson, S. C. (1987). *Politeness: Some universals in language usage.* Cambridge University Press.

10   Armaşu, V.-D. (2012). Modern approaches to Politeness Theory: A cultural context. *Lingua. Language and Culture, 1*, 9–19.

11   Goldsmith, D. J. (2007). Brown and Levinson's Politeness Theory. In B. B. Whaley & W. Samter (Eds.), *Explaining communication: Contemporary theories and exemplars.* Lawrence Erlbaum Associates, pp. 243–262.

12   Mao, L. R. (1994). Beyond Politeness Theory: "Face" revisited and renewed. *Journal of Pragmatics, 21*, 451–486. doi:10.1016/0378-2166(94)90025-6.

13   Venkataraman, B. (2024, May 30). What's the best way to fight viral disinformation? Look to South Florida. *The Washington Post.* www.washingtonpost.com/opinions/2024/05/30/misinformation-spanish-latinos-south-florida/

14   Cialdini, R. (2021). *Influence, new and expanded: The psychology of persuasion.* Harper Business.

15   Information Futures Lab (2024). *The information project: Rapid response pilot.* https://bpb-us-w2.wpmucdn.com/sites.brown.edu/dist/4/371/files/2024/01/Brown-IFL-Rapid-Reponse-Pilot-v3-5a656cc41c42028a.pdf

16   Starbird, K., Arif, A., & Wilson, T. (2024). *Disinformation as collaborative work: Surfacing the participatory nature of strategic information operations.* University of Washington preprint. https://faculty.washington.edu/kstarbi/Disinformation-as-Collaborative-Work-Authors-Version.pdf?utm_source=pocket_saves

17   Meyer, E. (2014). *The culture map: Breaking through the invisible boundaries of global business.* PublicAffairs/Hatchette.

18   Hall, E. T. (1976). *Beyond culture.* Anchor Books/Doubleday.

19   Hofstede, G. (1980). *Culture's consequences: International differences in work-related values.* Sage Publications.

20   Trompenaars, F., & Hampden-Turner, C. (2012). *Riding the waves of culture: Understanding diversity in global business* (3rd ed.). McGraw-Hill.

21   Nisbett, R. E. (2004). *The geography of thought: How Asians and Westerners think differently… and why.* Free Press.

22   Hofstede, G. (1980). *Culture's consequences: International differences in work-related values.* Sage Publications.

23 Follett, M. P. (1941/2003). In H. C. Metcalf & L. Urwick (Eds.), *Dynamic administration: The collected papers of Mary Parker Follett*. Routledge.

24 Hooker, J. (2012). Cultural differences in business communication. In C. B. Paulston, S. F. Kiesling, & E. S. Rangel (Eds.), *The handbook of intercultural discourse and communication*. Wiley, pp. 389–407.

25 Hammer, M. R. (2005). The Intercultural Conflict Style Inventory: A conceptual framework and measure of intercultural conflict resolution approaches. *International Journal of Intercultural Relations, 29*, 675–695. https://doi.org/10.1016/j.ijintrel.2005.08.010

26 Rahim, M. A. (1983). A measure of styles of handling interpersonal conflict. *Academy of Management Journal, 26*, 368–376.

27 Hall, E. T. (1976). *Beyond culture*. Anchor Books/Doubleday.

Chapter 9

Leading, Following, and Collaborating

Members of a team work together to achieve a common goal. Teamwork typically involves three kinds of roles: leading, following, and collaborating (which requires working together with other team members, under the leader). Our roles may change with context. Indeed, a manager is both a leader and follower—leading their team, but following bosses who are one or more levels above them.

At any given time, most of us probably won't be a leader, and may not even be a collaborator. Nevertheless, we still need to understand what people in the other roles are trying to accomplish so that we can contribute our part to a team effort. AIs are not likely to take over these roles, given that teams must often respond to open-ended situations that require taking context into account. Thus, it's important for all of us humans to understand leadership, followership, and human collaboration.

Leadership

Leadership is essential to any organization—large or small, national or local, for profit or not-for-profit. Any team without a leader runs the risk of being like an orchestra without a conductor, which soon becomes uncoordinated and beautiful music becomes discordant. Many researchers have studied what leaders do and what makes them successful, which we consider in the following sections.

What Leaders Do

To lead effectively, leaders must engage in a wide range of activities and do them well. Leaders must formulate a vision that defines their goals and provide compelling reasons why those goals are worth attaining. Once a goal is in place, leaders must develop a plausible strategy for how to attain it. If the leader works in a corporation, this may require them to assess market

DOI: 10.4324/9781032686653-11

trends, competitive landscapes, and internal capabilities to formulate strategies that give their organization an edge. But even leaders of non-profit organizations, neighborhood groups, and the like must create a plausible strategy for attaining their goals. After they create a strategy, leaders must define the specific tactics that they will use to implement it. Part of this process requires formulating a detailed agenda that defines specific tasks and sets priorities. Leaders must specify and order action items to create a detailed plan that is likely to achieve the goal. Following this, leaders must motivate the team to launch the plan, must delegate responsibilities skillfully, and need to allocate resources to allow each team member to carry out their tasks.

After initiating the plan, leaders cannot just sit back and watch as events unfold. Rather, they must manage their resources efficiently—including human, financial, and physical assets—as the plan is implemented. A leader needs to adjust tactics as the plan proceeds. As part of this, leaders must analyze situations, consider the input of others, weigh the pros and cons of different decisions, and then make informed choices, sometimes under considerable pressure. In addition, they often have to solve problems, or delegate this responsibility to others. In short, leaders need to engage in critical and creative thinking, need to manage their emotions, exercise emotional intelligence, and need to communicate effectively—as we reviewed in previous chapters.

Moreover, as time goes on, leaders must manage change within their organization. They must ensure that their organization adapts effectively to new challenges and opportunities. All the while, leaders are a primary custodian of the organization's culture, ethics, and values. They set the standard for integrity and ethical behavior and ensure that the organization's actions are aligned with its stated values.

And even more than this, leaders must regularly motivate followers to continue with them on the journey. It's not enough to motivate them to launch the plan—leaders need to inspire followers to hang in there over the long haul. To do so, leaders must articulate their vision in a way that captures followers' imaginations and aligns them with the organization's mission, vision, values, and immediate goals. To continue to motivate followers, leaders must understand individual and team dynamics, and they need to know how to recognize and reward contributions without causing resentment among other team members. They also must help team members keep in perspective the day-to-day work, seeing how it furthers the larger goals. Leaders typically rely on various motivational techniques to encourage, challenge, and support their team members (e.g.,[1,2,3,4,5]).

Leaders need to know how to help team members succeed in their tasks. In addition, in many organizations, leaders should help their team members grow and progress in their careers. Leaders may need to identify each team member's potential, be mentors, offer training opportunities, and give constructive feedback to facilitate personal and professional growth.

Leaders who are at the top of their organization often are also the face of their organization. They represent its interests in public forums, negotiations, and other external engagements. They both promote their organization and sense opportunities for joint ventures and collaborations. Leaders often network and develop relationships that benefit the organization.

Effective Leadership

Numerous researchers have studied leaders and formulated theories about how to be an effective leader.[6] Table 9.1 summarizes the most influential theories of leadership, namely Trait Theory,[7,8,9] Behavioral Theory,[10] Contingency Theory,[11,12] Transactional Theory,[13,14] Transformational Theory,[15,16] and Servant Theory.[17,18] The different theories often emphasize different aspects of what a leader must do, as just reviewed. Thus, a complete picture of effective leadership emerges when we consider the various theories and how they can complement each other.

Depending on what problem we are trying to solve, the guidelines provided by each leadership theory are more or less useful. In general, effective leadership requires us to be sensitive to the precise situation and adapt a blend of approaches to address the goal. And this is just the sort of circumstance where humans have an edge over an AI, given that the situation is open-ended (e.g., new factors emerge), we must take context into account, and must adapt in real-time to an evolving situation.

In spite of the fact that AIs are unlikely to serve as effective leaders in their own right, they can help human leaders hone their skills. Below is a prompt that can address two goals: First, this prompt can help us learn the different theories. Initially, we may often be guessing, but with appropriate feedback, over time we will come to identify the key aspects of each theory. Second, this prompt requires us to address the hard part: Figuring out how to combine elements from the different theories when they are appropriate. Although the AI might not be good at combining approaches and adjusting behavior in advance to fit the situation, it can give some useful feedback after the fact.

Table 9.1 Major Theories of Leadership.

Theory	Summary of Theory	Key Criticisms
Trait Theory	This theory posits that some people are born with or develop early in life certain traits that make them great leaders, such as intelligence, assertiveness, and extraversion.	The theory overlooks external factors such as situational context and interactions with followers; not all people who have the identified traits are successful leaders.
Behavioral Theory	Focusing on the behaviors of leaders, this theory categorizes them as task-oriented or people-oriented, and stresses the importance of balancing these behaviors.	It neglects the role of external environmental and situational factors.
Contingency Theory	According to this theory, success of leadership is contingent on adapting leadership style to the specific situation, such as the type of task and relationships with team members.	This theory is difficult to apply in practice; it lacks clear guidance on how to adapt to various situational factors.
Transactional Theory	This theory emphasizes that leaders should set clear goals and expectations; leaders should reward or punish followers based on their achievements. This is effective in stable environments.	This approach can inhibit innovation and cause frustration if goals are inconsistent or unexpected situations arise.
Transformational Theory	This theory advises leaders to inspire and motivate by transforming their followers' values and beliefs, aiming to induce them to exceed previous expectations.	This approach can undermine initiative and can neglect essential aspects of sound management.
Servant Theory	This theory emphasizes the leader's role as serving the greater good of the organization and supporting growth and fostering team members.	Following this advice carries the risk of becoming "leadership by following," which lacks long-term strategy or goal-driven decision making.

Me:

Please generate a scenario that describes a complex situation where leadership is required. When you generate the scenario, structure it to draw on two or more of the theories of leadership in the attached Table. Present the scenario to the user, and do not tell the user which theories you used when you created the scenario. After you present the scenario, pause and ask the user to describe what they would do if they were the leader, and to indicate which theories of leadership are guiding this response. Wait until the user has fully responded, and then provide feedback to the user. The feedback should mention the relevant aspects of the theories that would have been most appropriate to use (which are the ones you used to construct the scenario) and should evaluate the specific response. After you finish this, ask the user if they would like to do another scenario. Wait for the response. If the user says yes, please provide a new scenario and repeat the process. Attached is the table that summarizes the theories. [ATTACH TABLE 9.1]

AI:

Scenario:
Imagine you are the CEO of a technology start-up that is rapidly expanding in a highly competitive market. The company has grown from 10 employees to 100 in the last year. Recently, you've noticed a decline in productivity, and feedback indicates that employees feel disconnected from the company's mission and uncertain about their roles and responsibilities. There's also an increasing conflict between departments over resource allocation and priorities.

To address these issues, you decide to implement a comprehensive change in management strategy. This includes setting clear departmental goals and improving communication channels between teams. You also plan to introduce a series of workshops that aim to realign the employees with the company's core values and inspire them to contribute beyond their typical scope of work.

Questions:

1. What would you do if you were in this leadership position?
2. Which theories of leadership do you think are guiding your response?

Please take your time to think about this scenario and your approach, and share your thoughts. After you provide your response, I'll offer feedback on how it aligns with certain leadership theories and suggest improvements if necessary. Would you like to proceed with another scenario after this?

Me:

I would call an All Hands meeting and explain in detail why we need to pivot. My goal would be to draw on Transactional Leadership theory to motivate them, explaining how we absolutely need to achieve specific goals or we won't survive. I would also draw on Servant theory, telling them that I will want to meet with them weekly to talk through any general problems, and also that I will be more than happy to respond within 24 hours to emails on any issue.

AI:

Feedback on Leadership Approach
You've chosen a combination of leadership theories that are well-suited to the challenges described in the scenario. Your decision to call an All Hands meeting to address the entire company is a critical first step in reconnecting the employees with the company's mission and addressing the issues of role clarity and interdepartmental conflict.

1. **Transformational Leadership**: Your strategy to motivate employees by explaining the necessity of acquiring new skills and aligning them with the company's goals is a great example of Transformational Leadership. This theory focuses on inspiring and motivating employees to exceed their existing capabilities and to commit to the organization's vision. However, it seems

you mentioned drawing on Transactional Leadership in your response by mistake when describing the need for specific goal achievements, which slightly misaligns with your actual described actions that more closely fit the Transformational approach.

2. **Servant Leadership**: By offering to meet weekly with team members and promising a quick response time to emails, you are embodying Servant Leadership. This approach emphasizes supporting and growing each team member, and prioritizing their well-being and professional development, which can foster a strong, loyal team.

Suggestions for Improvement
While your approach is effective in many ways, here are a couple of enhancements you might consider:

• **Clarify Goals with Transactional Elements**: Even though your main approach aligns more with Transformational Leadership, incorporating clear, measurable goals from Transactional Leadership could further enhance team performance. Setting specific expectations and linking them to rewards can help ensure accountability and maintain a high level of productivity.

• **Incorporate Contingency Planning**: Considering the rapid growth and the dynamic market, employing aspects of Contingency Theory—by adapting your leadership style to the readiness and maturity of different teams or departments—might improve effectiveness. Tailoring your approach based on situational factors can be crucial during such scaling phases.

Overall, your leadership approach is thoughtful and well-structured for dealing with the scenario. Would you like to try another scenario to explore different leadership theories and responses?

Me:

No thank you.

AI:

Alright! If you have any other questions in the future or need further assistance, feel free to reach out. Have a great day!

Me:

Actually, I changed my mind. Yes, please, another scenario.

AI:

Scenario:
You are the principal of a large high school that has been underperforming in recent state assessments. Student engagement is low, teacher morale has declined, and there is a growing disconnect between the administration and the teaching staff. In response, you've decided to implement a new strategic initiative aimed at improving academic performance and school culture.

This initiative includes the introduction of a new curriculum focused on interdisciplinary learning, the establishment of a teacher mentorship program to enhance professional development, and the implementation of a more robust system for tracking student performance and engagement.

Questions:

1. What would you do if you were in this leadership position?
2. Which theories of leadership do you think are guiding your response?

Please take your time to think about this scenario and your approach, and share your thoughts. After you provide your response, I'll offer feedback on how it aligns with certain leadership theories and suggest improvements if necessary. Would you like to proceed with another scenario after this?

Me:

I would send out an inspirational memo and explain why the staff needs to participate and why it is good for them. I would do my best to draw on Transformation Leadership theory and explain why this is good for them.

AI:

Feedback on Leadership Approach
Your response indicates a clear intent to use Transformational Leadership, which is well-suited to the scenario of revitalizing a school environment where boosting morale and engagement is critical.

Transformational Leadership: By sending out an inspirational memo to explain the benefits of the new initiatives, you are focusing on motivating and aligning the staff with the vision for change. Transformational leaders inspire and motivate their followers to change expectations, perceptions, and motivations to work towards common goals. Your approach to explain why the changes are beneficial and how they align with the staff's personal and professional growth is a key component of this theory.

Suggestions for Improvement
While your initial approach sets a positive tone, here are a couple of enhancements you might consider to strengthen your strategy:

- **Incorporate Participative Leadership Elements**: Engage teachers and staff in the decision-making process about the new initiatives. This could include forming committees or holding meetings where they can voice their opinions and contribute ideas. This approach fosters a sense of ownership and commitment to the changes.
- **Utilize Servant Leadership to Support Transformational Goals**: Besides inspiring your staff, actively support their needs for professional development and personal growth. This can be achieved through the teacher mentorship program you mentioned. Make sure that these supports are clearly communicated and that you are responsive to the needs and challenges your staff might express.
- **Transactional Leadership for Structured Implementation**: While your focus on inspirational leadership is important, complementing it with aspects of Transactional Leadership, such as setting clear, measurable objectives for each

component of the initiative, could enhance accountability and clarity.

Your leadership approach has a strong foundation in transformational principles, which is crucial for driving significant cultural and performance changes in an educational setting. Would you like to try another scenario, or is there anything else you'd like to discuss?

Me:

No thank you.

AI:

Alright! If you need assistance in the future, feel free to reach out. Have a great day!

First, notice that as long as the thread was open, I could later pick up where I left off. I decided to continue because I wanted to provide a mediocre answer for the second scenario, to see what the AI would do. The AI was probably a bit too kind. My approach was ham-handed and likely to backfire. As discussed in Chapter 1, this sort of situation is veering into the realm where AIs are not adept. However, even so, this sort of exercise illustrates how an AI can familiarize the us with the theories and provide feedback about how we have applied them. If we squint in just the right way, we can view the fact that the feedback is not always quite right as a virtue: We should not accept such feedback at face value, but rather treat it as an impetus to think more deeply about the material. As we shall see in the following chapter, thinking through material is a crucial part of learning. Once again, the AI is serving as a spark for us to reflect and learn.

Moreover, we can use the "AI as a sounding board" approach more broadly. If we are faced with a tricky leadership decision, we can explain it to the AI and ask it to provide suggestions based on the different leadership theories. Alternatively, we can tell it what we plan to do and ask it to critique our plan based on the leadership theories.

Yes, this takes time—but many leaders already make the time to read "business books," recognizing that to stay on top of their game they need to continually learn. AIs offer a new way to learn, which they can personalize for each of our individual needs and preferences.

Followership

Most of us, most of the time, will not be leaders. Instead, we will be *followers* (e.g.,[19,20]). Even people who are leaders in one context may be followers in another. Although leaders are often in the forefront, followers are equally important. Without followers, nothing gets done. Effective followers have certain characteristics, which we can learn and hone. Again, we begin with the basics: What do followers do?

What Followers Do

Followers execute the leader's strategy and tactics, and provide feedback to help the team achieve its goal. Followers must not only support their leaders, but also augment what leaders do in the overall effort to achieve the team's goal. Followers should be engaged in the team's mission and contribute by carrying out their own roles well. Followers also need to work well with other team members. Followers should maintain positive relationships with all team members in the work context, putting aside personal differences—such as political views—that are not relevant to the job at hand.

In addition, followers should be reliable: They not only need to take responsibility for their assigned tasks, but they should also follow through on commitments and should respect deadlines. But more than this, followers should be proactive. They should think one step ahead and should be thinking both critically and creatively. Such thinking should lead them to ask questions and suggest alternative ways to reach the team's goal. At the same time, followers should be flexible and adaptable. They should note when circumstances change, and be able to retain focus on the team's goal even when the tactics or strategy change.

In order to accomplish these aims, followers need to communicate clearly and exercise emotional intelligence. In particular, they should listen well, derive a clear sense of a situation, and be comfortable sharing relevant information and concerns. Followers should also have the team's best interests in mind and should support decisions once they've been made. They also need to have integrity and uphold the team's values and objectives.

Finally, followers should be self-motivated, value personal growth, and manage their time effectively—perhaps with the help of an AI. Ideally, followers should enhance their personal and professional growth in ways that help the organization. For example, engaging in activities to enhance critical thinking and creative problem solving can build a solid foundation for being able to take on new challenges and innovative roles in an organization.

Table 9.2 Key Theories of Followership.

Theory	Summary of Theory	Key Criticisms
Five Type Theory	This theory categorizes followers along two dimensions: critical thinking and active engagement. Based on different combinations of values along these dimensions, it defines five types of followers: Star, Conformist, Passive, Alienated, and Pragmatic Survivors. The theory highlights how personality and skills contribute to how effective followers are.	Critics argue that the theory oversimplifies the impact of personality and skills on how effective followers are, failing to account for the complexity of personality traits and their situational effects.
Courageous Follower Model	This model emphasizes the importance of courage for effective followership, necessary both to support and to challenge leaders. It stresses the need to find a balance between following and questioning authority.	This theory has been criticized as idealistic; it assumes a level of ethical certainty and personal risk-taking that may not be practical or culturally consistent for many employees.
Theory of Independent Followership	This theory argues that followers should be empowered to function independently without strong leadership, similar to how string quartets operate without a conductor.	This theory has been criticized for understating the importance of leadership, particularly in complex tasks, and overstating the ability of followers to self-organize.
Social Exchange Theory	This theory proposes that effective followership is based on reciprocal, trust-filled relationships that benefit all involved. It relates to the Synergistic Followership Model, which highlights the mutual enhancement of strengths among leaders and followers.	Critics argue that this theory may overemphasize the transactional aspects of relationships and is overly optimistic about the level of collaboration that can be achieved in teams.
Trait Theory	According to this theory, specific personality traits, such as high Conscientiousness, high Openness, high Agreeableness, and low Neuroticism make a person a good follower. Research shows that these traits do correlate with effective teamwork.	Researchers argue that this theory oversimplifies human behavior, neglecting the dynamic nature of interactions and situational factors (e.g., nature of the task, resources) affecting group dynamics.
Behavioral Theory	This theory focuses on specific behaviors such as effective communication, careful listening, and adaptability as keys to effective followership. It has been supported by studies linking these behaviors to team success.	Critics have argued that this theory does not address why people exhibit these behaviors and does not analyze underlying traits or situational factors (such as organizational culture or team dynamics).

Effective Followership

Numerous theories and models specify how followers can be most effective. Table 9.2 summarizes the major theories and models, and summarizes key criticisms of them. The theories are Five Type Theory,[21,22,23] Courageous Follower Model,[24,25] Theory of Independent Followership,[26,27] Social Exchange Theory,[28,29,30,31,32] Trait Theory,[33,34] and Behavioral Theory.[35,36,37]

Each of these theories and models highlights different aspects of what makes someone an effective follower. These theories again center on human interactions in open-ended situations that require taking context into account, which are likely to challenge current AIs. Nevertheless, I found it useful to have GPT-4 simplify the take-home message from the different theories. I asked the following, attaching Table 9.2, and it responded as below:

Me:

Please extract the key strengths of each of the theories summarized in the attached table, and develop a new theory. The new theory should not just be a list of these strengths, but rather a novel synthesis. [ATTACH TABLE 9.2]

AI:

Based on the key strengths of the theories summarized in the attached table, I've extracted the following main strengths for each:

1. **Five Type Theory**: This theory highlights the importance of personality and skills in determining follower effectiveness, suggesting that different traits lead to different types of followership.
2. **Courageous Follower Model**: Emphasizes courage as a crucial component for followers, advocating for a balance between supporting and challenging authority, which is key for ethical and dynamic followership.
3. **Theory of Independent Followership**: Advocates for the empowerment of followers to operate independently, highlighting the potential for self-organization similar to non-hierarchical structures like string quartets.

4. **Social Exchange Theory**: Focuses on the reciprocal, trust-filled relationships that can enhance collaborative efforts and mutual benefits among followers and leaders.
5. **Trait Theory**: Identifies specific personality traits that correlate with effective followership and teamwork, providing a clear framework for understanding follower dynamics.
6. **Behavioral Theory**: Concentrates on observable behaviors such as effective communication and adaptability, linking these directly to successful team outcomes.

Proposed New Theory: Integrated Dynamics of Followership (IDF)

The "Integrated Dynamics of Followership" (IDF) theory synthesizes the strengths of existing followership theories into a comprehensive framework that emphasizes the dynamic interplay between follower traits, behaviors, and situational contexts. This theory posits that effective followership results from the following core components:

1. **Adaptive Trait-Behavior Nexus**: IDF integrates Trait and Behavioral Theories by suggesting that effective followership behaviors emerge from a combination of inherent personality traits and situational adaptations. Followers are not only influenced by their stable traits but also by the need to adapt their behaviors based on team needs and environmental cues.
2. **Situational and Relational Dynamics**: Drawing from the Social Exchange and Independent Followership Theories, IDF emphasizes that the effectiveness of followership varies with the situational context, including the nature of the task and the relational dynamics within the team. This component recognizes the importance of empowerment and the potential for followers to lead initiatives in the absence of formal leadership, fostering a flexible, self-organizing team environment.
3. **Ethical Courage and Engagement Spectrum**: Inspired by the Courageous Follower Model and the Five Type Theory, IDF suggests a spectrum of follower engagement, from passive to active, where the optimal point is influenced by the follower's courage to act ethically and constructively. This spectrum helps followers navigate the balance between obedience and autonomy,

encouraging them to adapt their level of engagement based on ethical considerations and the demands of the situation.

By integrating these components, the IDF offers a flexible yet structured approach to understanding and improving followership across different organizational contexts. It acknowledges the complex interdependencies between follower characteristics, behaviors, and the external environment, advocating for a holistic view of followership as a dynamic and influential role within teams and organizations.

Me:

How does this new theory address the main criticisms of the previous theories?

AI:

The Integrated Dynamics of Followership (IDF) theory addresses the main criticisms of the previous theories by creating a more holistic and adaptable framework for understanding followership. Here's how it responds to each criticism:

1. **Oversimplification of Traits and Behaviors:**

   o **Criticism:** The Five Type Theory and Trait Theory have been criticized for oversimplifying how personality traits impact follower effectiveness, and the Behavioral Theory for not addressing the underlying reasons behind specific behaviors.

   o **IDF's Response:** By integrating traits and behaviors into a single, adaptive trait-behavior nexus, IDF acknowledges that effective followership behaviors emerge not only from stable personality traits but also from the need to adapt behaviors to specific situational demands. This approach recognizes the dynamic nature of interactions and the complexities involved in real-world environments.

2. **Practicality and Ethical Concerns:**

   o **Criticism:** The Courageous Follower Model has been deemed idealistic for assuming a level of ethical certainty and risk-taking that may not be universally applicable or practical.

o **IDF's Response**: IDF introduces the concept of an ethical courage and engagement spectrum, which allows followers to adjust their level of engagement and courage based on situational ethics and practicality. This flexibility ensures that followers can maintain ethical integrity without compromising on practical concerns in varying cultural and organizational contexts.

3. **Underestimating Leadership and Overstating Self-organization**:

   o **Criticism**: The Theory of Independent Followership has been critiqued for understating the importance of leadership and overstating the capacity for self-organization among followers.
   o **IDF's Response**: IDF emphasizes situational and relational dynamics, which include the recognition of when leadership is necessary and when followers can effectively self-organize. This balanced approach allows for adaptive leadership and follower roles, depending on the complexity of the task and the maturity of the team.

4. **Overemphasis on Transactional Relationships**:

   o **Criticism**: Social Exchange Theory has been criticized for focusing too much on the transactional aspects of relationships, potentially overlooking the intrinsic motivations and the non-transactional elements that can enhance teamwork.
   o **IDF's Response**: By incorporating a broader view of reciprocal, trust-filled relationships that go beyond mere transactions, IDF supports a more nuanced understanding of collaboration. It highlights mutual benefits while also considering the intrinsic motivations that drive followers and leaders to contribute to a collective goal.

In essence, the IDF theory seeks to provide a more nuanced, realistic, and applicable approach to followership by addressing the weaknesses of existing theories through an integrated and contextually aware framework. This comprehensive approach ensures that followership is viewed as a multifaceted and dynamic role crucial to the success of any organization.

The AI was creative in how it integrated the theories, and did so in a way that is potentially useful. However, its proposed theory is vague and abstract. It is not clear how this theory would be fleshed out, and the potential utility of this theory depends on these details. Without the details, it's more of an approach—as it sometimes notes—than an actual theory. Clearly, being a good follower depends on managing open-ended situations that require taking context into account. And, again, AIs can assist us only so far in grappling with these sorts of interactions. Nevertheless, we can take the same approach we took with leadership, and have the AI create activities and exercises that help us learn essential aspects of the underlying skills.

Collaboration

Being an effective team member typically relies on more than just following a leader: We must also work closely with colleagues. "Collaboration" occurs when two or more team members, or groups, work together to achieve a shared goal. Not all teamwork requires collaboration. In some cases, teamwork merely requires members to work independently on separate components of the overall tasks. In contrast, collaboration requires team members to interact as they work together.

What Do Collaborators Do?

People who collaborate on a team engage in joint activities that help to achieve the team's goal. First and foremost, collaborators must communicate, both with each other and with the leader of the team. Such communication occurs in a variety of modalities, including presentations and discussions during meetings (both in person and virtual), email, texts, and via specialized project management tools (such as Asana or Slack). Collaborators not only discuss progress and plans, but also provide feedback on performance, identify challenges, and suggest ways to move ahead. Moreover, in some teams, team members must make tactical decisions, and need to communicate with stakeholders to ensure that they contribute to such decisions.

In addition, collaboration requires team members to devise ways to work together to carry out their specific tasks. For example, they must coordinate—and sometimes negotiate—their schedules and roles. Team members also must coordinate various other tasks that must be completed to reach the goal, such as writing progress reports, which may require using shared documents or platforms for providing feedback and sharing editing.

Effective Collaboration

To collaborate well, team members must trust each other. Trust is built up over time, when team members realize that they can count on each other to follow through on commitments, to do a good job, to be reliable, and to communicate effectively. Trust requires a degree of risk.[38,39] One person must be willing to be vulnerable to another, expecting that the other person will follow through as expected.

Another characteristic of successful collaboration flows from trust, namely that team members communicate effectively. This requires sharing information freely when appropriate, listening closely, and working to ensure that all parties understand each other. Team members must provide accurate and actionable feedback to each other, which requires them to be sensitive to non-verbal cues.[40] A slight grimace or hint of a smile can often be worth many words.

Mutual respect is another characteristic of effective collaborators. Respect requires becoming familiar with another person's strengths, and learning how they can help achieve the shared goal. These strengths can focus on the person's character, skills, and/or abilities. "Character" includes attributes such as their work ethic and personality traits, such as conscientiousness. "Skills" include both general skills, such as those that underly critical thinking and creative problem solving, and specific skills that are relevant for the task at hand. And "abilities" include general intelligence and specific "natural inclinations," such as mechanical ability.

Another key factor for successful collaboration is having a healthy organizational culture. For example, the organizational culture should allow team members to admit mistakes and not to fear retribution when they ask for help. Nobody should be embarrassed to have tried something new that didn't work out, provided that they were thoughtful and serious. An organizational culture need not develop haphazardly. When creating a team, the initial phases of interaction can establish a culture by focusing on the "rules of the road," providing clear guidelines and expectations—which everyone agrees to follow (cf. [41]).

We can employ the Cognitive Amplifier Loop (CAL) to become familiar with and apply collaboration skills. Indeed, we can simply adapt the prompt used earlier to become familiar with leadership skills. But, once again, an AI can only help us go so far, given the fact that collaborations often occur in open-ended situations that require taking context into account. However, it is still useful to become familiar with the essentials discussed above, and hence worth the time and trouble to adapt the previous prompt to help people learn collaboration skills.

In addition, let's consider a particular skill that is essential for effective collaborations, namely the capacity to resolve conflicts. Even in the most productive teams, leaders, followers and team members will not always agree. Researchers and practitioners have formulated many principles of effective conflict resolution (e.g.,[42,43,44]). Because there are many principles, it's difficult to memorize them all, let alone master how to apply them appropriately. An AI can help us learn these principles and learn to apply them.

In what follows, the AI engages a user in a role-playing game that is designed to help them become familiar with principles of conflict resolution and learn how best to apply them. I derived the particular principles incorporated in this exercise from the references cited above, but we can substitute specific principles that are relevant to a particular situation.

Me:

1. Your task is to develop complaints for a role-playing game and then use them to train people how to use the principles of conflict resolution effectively.

2. To train the user, please engage in a role-playing game. You will play the role of a disgruntled person and the user has to interact with you, trying to use principles of conflict resolution to remain safe, avoid ethical or legal issues, and calm you down and resolve the problem.

3. Please develop the conflicts so that they all occur in some aspect of a medical practice. Here is an example of such a conflict in an oncology clinic: You state, with a raised voice, "I am extremely upset about my treatment plan! The information the clinic gave me wasn't what actually happened, and now I don't trust the care team. I need the oncologist and nursing staff to explain to me what happened, and tell me exactly what they will do differently in the future, so this won't happen again!" This situation is similar to a customer complaint in other service industries. Every member of the care team (the nurses, physicians, charge nurse, and manager) must have good conflict resolution skills to identify the problem and then work with the person who is unhappy with the situation.

4. After you present a complaint from your point of view, ask the user to respond to your complaint. Tell the user "Please respond in detail as if you were in this situation. Take your time to

formulate your response, focusing on using the principles of conflict resolution. After you have formulated your response, please proceed to type it out. Once you have provided your detailed approach, we will move on to the feedback stage."

5. The user should then input their response. Because this is a text-based interaction, the user's act of typing and submitting their response serves as the "pause" and "input" in this training session. Wait for the user to respond before you continue. Do not proceed until the user has fully responded. Only after the user has fully responded should you continue.

6. After the user's response is received, only then do you move on to provide feedback. Clearly state that the feedback will be based on the user's submitted response, focusing on how well they applied the principles of conflict resolution and suggesting areas for improvement. Base your feedback on the following three categories of principles of conflict resolution in a medical context:

   6.1. Physical Safety: Never, under any circumstances, allow an interaction to escalate to the point where any person is in physical danger. If the patient shows any signs of being violent, remove yourself from the situation as quickly as possible.

   6.2. Ethical and Legal Vulnerability: Please include in your feedback an assessment of how well the response adheres to legal and ethical standards. Not only should the response be honest and ethical, but it should not leave the clinic open to a lawsuit. Do not admit fault or take responsibility for any action that could lead to a lawsuit. Do not apologize, do not make promises about future behavior that you may not be able to fulfill. When providing feedback, please assess whether the response inadvertently admits fault or makes commitments that exceed professional boundaries or capabilities, offering alternative phrasings or approaches where necessary. Ask for constructive alternatives when responses might lead to legal or ethical issues. If a response contains potential legal or ethical issues, please suggest how it could be rephrased to maintain empathy and support while also protecting against legal risks.

6.3. De-escalation. Respecting the Physical Safety and Ethical and Legal Vulnerability principles, use the following principles to de-escalate the conflict: a) Active Listening: Engaging in active listening involves paying close attention to what the other party is saying, acknowledging their points, and demonstrating understanding. b) Empathy: Trying to understand and empathize with the other person's feelings and viewpoints can help de-escalate tensions and foster a cooperative environment. Empathy involves seeing the conflict from the other person's perspective and acknowledging their emotions. c) Non-Confrontational Communication: Using "I" statements instead of "you" statements helps in expressing one's own feelings and perspectives without blaming or accusing the other party. This type of communication reduces defensiveness and promotes open dialogue. d) Finding Common Ground: Identifying areas of agreement, even if they are minimal, can provide a foundation for building a mutually acceptable solution. Focusing on shared interests or goals helps shift the interaction from adversarial to collaborative. e) Problem-Solving: Instead of focusing on past conflicts or assigning blame, effective conflict resolution emphasizes identifying the underlying issues and working together to find creative and mutually satisfying solutions. f) Compromise and Flexibility: Being willing to compromise and show flexibility can be essential for reaching a resolution. This might involve each party giving up something or agreeing to a middle-ground solution. g) Establishing Clear Agreements: Once parties have reached a resolution, it's important to articulate the agreement clearly, specifying any agreed-upon actions and commitments. h) Neutrality and Impartiality: In situations where a third party, like a mediator, is involved, it's crucial that this person remains neutral and impartial, facilitating the conversation without taking sides. i) Respect and Dignity: Maintaining respect for each other throughout the conflict resolution process is vital. Even when disagreements are intense, acknowledging each other's inherent dignity can preserve relationships and lead to more enduring solutions.

7. Remember: After you present a complaint, ask the user to respond as if they were in that situation. Then pause. Do not continue. Do not provide feedback until the user has fully responded. Only after the user has fully responded should you continue. After you have waited for the response and the user has responded, at that point provide feedback about the principles that were used well or need additional attention, based on the user's response.

8. After you provide feedback, end with an interactive closure, asking whether the user would like to engage with another scenario, thus providing a clear end to one interaction loop and a potential beginning for another. If they say yes, please provide one and repeat the above process. If they say no, thank them in an encouraging way and stop the training.

AI:

Let's begin with a scenario set in a medical practice, specifically a pediatric clinic.

Complaint:
You are a healthcare professional at a pediatric clinic. I, as the disgruntled parent, come to you and say in a visibly upset tone, "I'm very frustrated and disappointed! I was told my child would receive a comprehensive developmental assessment during our visit, but the doctor barely spent any time with us and seemed rushed. I don't feel like my child's needs are being taken seriously, and I'm worried crucial issues might be missed or dismissed. I need assurance that my child's health and development are prioritized, and I want a clear explanation of how the clinic will address my concerns to prevent this from happening again."

Your task:
Please respond in detail as if you were the healthcare professional in this situation. Take your time to formulate your response, focusing on using the principles of conflict resolution. After you have formulated your response, please proceed to type it out. Once you have provided your detailed approach, we will move on to the feedback stage.

Me:

"I'm sorry that you are upset. I will look into what the doctor did, and why they took so little time. Because the doctor is very experienced and familiar with a wide range of developmental issues, I suspect that they were quickly able to zero in on the problem and have a very good idea of an effective treatment plan. I assure you that we are taking your child's needs very seriously. Let me talk to them, and I will get back to you."

AI:

Thank you for your response. Now I'll provide feedback based on the principles of conflict resolution in a medical context:

Physical Safety: Your response did not escalate the situation and
   maintained a safe environment, which is commendable.

Ethical and Legal Vulnerability:

   • Acknowledging the parent's concerns and promising to inves-
      tigate is positive, as it demonstrates responsiveness and
      responsibility. However, it's important to be cautious with the
      language that implies assurance before all facts are gathered.
      While you did not explicitly admit fault, stating you will "look
      into what the doctor did" could be perceived as presupposing
      an error, which might create a vulnerability.
   • Your explanation attempting to justify the doctor's behavior
      ("the doctor was quickly able to zero in on the problem") is
      a double-edged sword. It's good to provide reasons for the
      doctor's actions, which can help in de-escalating the situation.
      However, this should be framed as a hypothesis rather than a
      definitive conclusion until all information is reviewed, to avoid
      potential misunderstandings or false assurances.
   • Ensure that any promises made (like getting back to the parent)
      are followed through to avoid creating expectations that might
      not be met, which can lead to further issues and a loss of trust.

De-escalation:

   • Active Listening: You demonstrated this by acknowledging the
      parent's feelings of frustration and disappointment.

- Empathy: Your response showed an attempt to empathize by acknowledging the seriousness with which the clinic takes the child's needs. However, further empathetic statements, such as recognizing the parent's anxiety over their child's health, could enhance the effectiveness.
- Non-Confrontational Communication: Your use of "I" statements was appropriate and helped present the message without being confrontational.
- Finding Common Ground: Affirming that the child's needs are a priority helps establish shared concerns.
- Problem-Solving and Compromise: You indicated a willingness to investigate and follow up, which are constructive steps toward resolving the issue.
- Clear Agreements: You concluded with a clear action item (talking to the doctor and getting back to the parent), which is good. Ensure this is carried out promptly.
- Respect and Dignity: Maintaining a respectful tone throughout your response was effective in honoring the parent's dignity and fostering a more collaborative interaction.

In conclusion, your response was well-handled in many aspects but could be further improved by being careful with the assumptions and ensuring all statements are verifiable. It's also vital to reinforce empathy explicitly by acknowledging the parent's stress and worry in more depth.

Would you like to engage with another scenario?

Me:

Yes, please.

AI:

Let's move on to the next scenario, which is set in a cardiology clinic.

Complaint:
You are a nurse in a cardiology clinic. I, as the patient, approach you with noticeable frustration and say, "This is unacceptable! I've

been waiting for over an hour past my scheduled appointment time to see the cardiologist. Every time I come here, there's a delay, and nobody seems to care about my time. I have other commitments and cannot afford to spend half my day here. I need to understand why these delays keep happening and what will be done to ensure that my future appointments are on time. I expect better service, especially when it concerns my heart health!"

Your task:
Please respond in detail as if you were the nurse in this situation. Focus on using the principles of conflict resolution to address the patient's concerns. After you have formulated your response, please type it out. Once you have provided your detailed approach, we will move on to the feedback stage.

Me:

"I understand how frustrating this is, and please do know that we value your time and your faith in us as a patient. The doctors here are excellent, and tend to spend a lot of time with each patient, trying to understand them and help them. I don't know why this is a chronic problem, and will look into how long the delays are in general. If they are in fact typically long, I will take this up with my supervisor and see what we can do to address the problem. In the meantime, let me look into the situation right now, and see what we can do to get you seen soon."

AI:

Thank you for your response. Now, I'll provide feedback based on the conflict resolution principles:

Physical Safety: You maintained a calm and respectful demeanor, which helps ensure a safe environment free from escalation.

Ethical and Legal Vulnerability:

- Your acknowledgment of the patient's frustration and assurance of the clinic's care for their time are positive steps. However, it's crucial to avoid making broad promises about systemic changes without having the authority or specific plan to implement such changes.

- Investigating the cause of the delays and discussing them with your supervisor are appropriate steps. However, ensure any commitments made are within your capacity to fulfill to avoid potential issues of accountability or expectations that are not met.
- Clarify that while you aim to address their concerns, some aspects of healthcare, like the time doctors spend with patients, are inherently unpredictable and focused on patient care quality.

De-escalation:

- Active Listening: You demonstrated active listening by acknowledging the patient's distress and understanding their perspective.
- Empathy: Your empathetic response, recognizing the inconvenience and frustration, helps validate the patient's feelings, which is crucial for de-escalation.
- Non-Confrontational Communication: Using "I" statements effectively expressed your intentions without attributing blame or creating defensiveness.
- Finding Common Ground: You emphasized the clinic's commitment to patient care, which aligns with the patient's expectations, thus finding common ground.
- Problem-Solving: You offered immediate and future-oriented solutions by proposing to check the current status and addressing broader systemic issues with your supervisor.
- Compromise and Flexibility: Indicating willingness to look into the issue and seek resolutions shows a readiness to find a middle ground.
- Clear Agreements: Committing to check the current delay and inform the patient shows a clear next step. Ensure that this action is taken promptly and communicated effectively.
- Respect and Dignity: You maintained a respectful tone throughout, which helps preserve the patient's dignity and fosters a more collaborative interaction.

Overall, your response was considerate and constructive, focusing on immediate and potential long-term solutions. Just ensure that any commitments are actionable and within your scope, and maintain transparent communication about what can realistically be achieved.

Would you like to try another scenario?

Me:

No, thank you. I look forward to another session with you later!

AI:

You're welcome! I'm glad to hear you found the session helpful. Whenever you're ready for another session or have any other questions or topics to explore, feel free to reach out. Have a great day!

Clearly, the CAL can help people learn to apply these sorts of principles. Note, however, that although the AI can effectively teach the principles, there are limits to the quality of the feedback an AI can provide about human interactions. If the scenario involves nuances that hinge on the specific context, the AI may not be able to take those into account. Nevertheless, we can get reasonably far by using an AI to help with basic training. As usual, the AI provides immediate, personalized feedback, which is difficult to obtain in traditional educational settings—and impossible to receive from reading a book.

Flourishing with the CAL

Leading, following, and collaborating are human activities. Because AIs don't have their own goals or motivations, they can't really assume any of these roles. However, as we've seen here, the CAL can in fact help us to hone our skills and knowledge in these areas, building on strengths and helping us address limitations. Learning to work well with others is a key aspect of feeling satisfied at work, which can lead to having enough money and having a good work–life balance. In addition, building these skills requires personal growth and developing our abilities and talents, be they as leaders, followers, and/or collaborators. Working with others to achieve a worthy common goal can also provide a sense of purpose or meaning in life.

We next turn to Part III of this book, which addresses essential skills and knowledge we all need to master if we are to adapt to the changes that are on the way. The sea change we are now facing is not a one-time event, and will not be over quickly. We now live in an age of rolling AI thunder, and we need to embrace it and appreciate the good it can bring but also be alert to the problems it may present. Meeting these challenges requires not only learning how to learn effectively and efficiently, but also adopting a mindset of lifelong learning. Moreover, many of us may need to reconsider, or adapt, the big-picture goals and perspectives that confer meaning in our lives.

Digging Deeper

The following videos provide accessible, in-depth treatments of different aspects of the material discussed in this chapter. As usual, the search terms can help to locate new videos on the topic.

Simon Sinek: Inspiring Others To Do Remarkable Things: https://www.youtube.com/watch?v=C5vqyKRJzEI
   The puzzle of motivation | Dan Pink | TED: https://www.youtube.com/watch?v=rrkrvAUbU9Y

- "Leadership roles and responsibilities"
- "Setting goals and devising strategies as a leader"
- "Motivating and inspiring followers"
- "Managing resources and delegating responsibilities"
- "Adapting to change as a leader"
- "Promoting organizational culture and values"
- "Developing and mentoring team members"
- "Representing the organization as a leader"
- "Critical thinking and problem solving in leadership"
- "Emotional intelligence in leadership"
- "Effective communication strategies for leaders"
- "Collaboration and networking as a leader"

Theories of Leadership: https://www.youtube.com/watch?v=Ifd6jxpVx6k
   Servant Leadership: https://www.youtube.com/watch?v=H8rdRFVG0rE

- "Trait theory of leadership"
- "Behavioral theory of leadership"
- "Contingency theory of leadership"
- "Transactional theory of leadership"

- "Transformational theory of leadership"
- "Servant theory of leadership"
- "Comparing leadership theories"
- "Situational factors in leadership effectiveness"
- "Adapting leadership styles to context"
- "AI-assisted leadership training"
- "Scenario-based leadership development"
- "Applying leadership theories in practice"

The Art of Followership in Leadership: https://www.youtube.com/watch?v=XBL6oTaGu-w
  Being an Effective Follower: https://www.youtube.com/watch?v=wddfzAO--VQ

- "Characteristics of effective followers"
- "Importance of followership in organizations"
- "Supporting leaders as a follower"
- "Proactive followership strategies"
- "Flexibility and adaptability in followership"
- "Communication skills for effective followership"
- "Loyalty and integrity in followership"
- "Maintaining positive team relationships as a follower"
- "Self-management and personal growth in followership"
- "Enhancing critical thinking and problem-solving skills as a follower"
- "Followership and organizational success"
- "Balancing followership and leadership roles"

Leadership Perspectives Webinar—Followership with Barbara Kellerman: https://www.youtube.com/watch?v=31YChjPoPn8
  Types of Followers: https://www.youtube.com/watch?v=AtbnTNld-LM

- "Robert Kelley's Five Type Theory of followership"
- "Ira Chaleff's Courageous Follower Model"
- "Abraham Zaleznik's Theory of Independent Followership"
- "Social Exchange Theory in followership"
- "Synergistic Followership Model"
- "Trait Theory of effective followership"
- "Behavioral Theory of effective followership"
- "Criticisms of followership theories"
- "Situational factors in followership effectiveness"
- "Personality traits and effective followership"
- "Behaviors of effective followers"
- "Integrating followership theories and models"

Shane Snow helps us to Build Dream Teams: https://www.youtube.com/watch?v=FVNC2Nd18xk

Conflict Resolution Training: How To Manage Team Conflict In Under 6 Minutes! https://www.youtube.com/watch?v=PHJ8eybXJdw

- "Characteristics of effective collaboration"
- "Building trust in team collaboration"
- "Effective communication strategies for collaboration"
- "Fostering a culture of collaboration in teams"
- "Mutual respect and collaboration"
- "AI-assisted conflict resolution training"
- "Principles of effective conflict resolution"
- "De-escalation techniques in conflict resolution"
- "Ethical and legal considerations in conflict resolution"
- "Role-playing exercises for conflict resolution"
- "Lifelong learning and collaboration skills"

References

1   Deci, E. L., & Ryan, R. M. (1985). *Intrinsic motivation and self-determination in human behavior*. Plenum.

2   Deci, E. L., & Ryan, R. M. (2000). The "what" and "why" of goal pursuits: Human needs and the self-determination of behavior. *Psychological Inquiry, 11*, 227–268. https://doi.org/10.1207/S15327965PLI1104_01

3   Ryan, R. M., & Deci, E. L. (2000). Self-determination theory and the facilitation of intrinsic motivation, social development, and well-being. *American Psychologist, 55*, 68–78. doi:10.1037//0003-066x.55.1.68.

4   Vroom, V. H. (1964). *Work and motivation*. Wiley.

5   Vroom, V. H. (2005). On the origins of expectancy theory. In K. G. Smith & M. A. Hitt (Eds.), *Great minds in management*. Oxford University Press, pp. 239–258. https://doi.org/10.1093/oso/9780199276813.003.0012

6   Northouse, P. G. (2021). *Leadership: Theory and practice* (9th ed.). Sage.

7   Colbert, A. E., Judge, T. A., Choi, D., & Wang, G. (2012). Assessing the trait theory of leadership using self and observer ratings of personality: The mediating role of contributions to group success. *The Leadership Quarterly, 23*, 670–685. https://doi.org/10.1016/j.leaqua.2012.03.004

8   Judge, T. A., Bono, J. E., Ilies, R., & Gerhardt, M. W. (2002). Personality and leadership: A qualitative and quantitative review. *Journal of Applied Psychology, 87*, 765–780. doi:10.1037/0021-9010.87.4.765.

9   Zhang, J., Yin, K., & Li, S. (2022). Leader extraversion and team performance: A moderated mediation model. *PLoS One, 17*(12), e0278769. doi:10.1371/journal.pone.0278769.

10 Humphrey, R. H. (2013). *Effective leadership: Theory, cases, and applications*. Sage Publications.

11   Deckard, G. J. (2009). Contingency theories of leadership. *Organizational behavior in health care, 2*, 191–208.

12   Hughes, R. L., Ginnett, R. C., & Curphy, G. J. (2022). *Leadership: Enhancing the lessons of experience* (10th ed.). McGraw-Hill.

13   Aga, D. A. (2016). Transactional leadership and project success: The moderating role of goal clarity. *Procedia Computer Science, 100*, 517–525.

14   Tavanti, M. (2008). Transactional leadership. In A. Marturano & J. Gosling (Eds.), *Leadership: The key concepts*. Routledge, pp. 166–170.

15   Bass, B. M., & Riggio, R. E. (2006). *Transformational leadership* (2nd ed.). Lawrence Erlbaum Associates.

16   Hay, I. (2006). Transformational leadership: Characteristics and criticisms. *E-Journal of Organizational Learning and Leadership, 5*(2). www.academia.edu/9121644/Transformational_leadership_characteristics_and_criticisms

17   Berger, T. A. (2014). Servant leadership 2.0: A call for strong theory. *Sociological Viewpoints, 30*, 146–167.

18   Van Dierendonck, D., & Patterson, K. (2010). Servant leadership: An introduction. In D. van Dierendonck & K. Patterson (Eds.), *Servant leadership: Developments in theory and research*. Palgrave Macmillan UK, pp. 3–10.

19   Matshoba-Ramuedzisi, T., de Jongh, D., & Fourie, W. (2022). Followership: A review of current and emerging research. *Leadership & Organization Development Journal, 43*, 653–668. https://doi.org/10.1108/LODJ-10-2021-0473

20   Uhl-Bien, M., Riggio, R. E., Lowe, K. B., & Carsten, M. K. (2014). Followership theory: A review and research agenda. *The Leadership Quarterly, 25*, 83–104. https://doi.org/10.1016/j.leaqua.2013.11.007

21   Kelley, R. (1992). *The power of followership*. Doubleday Business.

22   Kelley, R. (2008). Rethinking followership. In R. Riggio, I. Chaleff, & J. Lipman-Blument (Eds.), *The art of followership: How great followers create great leaders and organizations*. Wiley, pp. 5–17.

23   Ligon, K. V., Stoltz, K. B., Rowell, R. K., & Lewis, V. J. (2019). An empirical investigation of the Kelley Followership Questionnaire Revised, *Journal of Leadership Education, 18*, 97–112.

24   Chaleff, I. (2009). *The courageous follower: Standing up to and for our leaders* (3rd ed.). Berrett-Koehler Publishers.

25   Yuankun, C., Qing, Z., Shanshi, L, & Zhenbing, Z. (2019). The courageous followership behavior: A literature review and prospects. *Foreign Economics & Management, 41*, 47–60.

26   Zaleznik, A. (1965). *The dynamics of subordinacy*. Harvard Business Review.

27   Kellerman, B. (2016). Leadership—it's a *system*, not a person! *Daedalus, 145*, 83–94. www.amacad.org/publication/leadership-its-system-not-person

28   Baker, S. D. (2007). Followership: The theoretical foundation of a contemporary construct. *Journal of Leadership & Organizational Studies, 14*, 50–60. https://psycnet.apa.org/doi/10.1177/0002831207304343

29   Losada, M., & Heaphy, E. (2004). The role of positivity and connectivity in the performance of business teams: A nonlinear dynamics model. *American Behavioral Scientist, 47*, 740–765. https://doi.org/10.1177/000276420 3260208

30   Zou, W. C., Tian, Q., & Liu, J. (2015). Servant leadership, social exchange relationships, and follower's helping behavior: Positive reciprocity belief matters. *International Journal of Hospitality Management, 51*, 147–156. https://doi.org/10.1016/j.ijhm.2015.08.012

31   Cropanzano, R., Anthony, E. L., Daniels, S. R., & Hall, A. V. (2016). Social Exchange Theory: A critical review with theoretical remedies. *The Academy of Management Annals, 11*, 1–38. https://doi.org/10.5465/annals.2015.0099

32   Zafirovski, M. (2005). Social Exchange Theory under scrutiny: A positive critique of its economic-behaviorist formulations. *Electronic Journal of Sociology.* https://epe.lac-bac.gc.ca/100/202/300/ejofsociology/2006/11/content/2005/tier2/SETheory.pdf

33   Kudek, D. R., Winston, B., & Wood, J. A. (2020). Followership and the relationship between Kelley's followership styles and the Big Five factor model of personality. *Journal of Organizational Psychology, 20*, 102–117. https://doi.org/10.33423/jop.v20i3.2943

34   Utomo, T., Handoyo, S., & Fajrianthi, F. (2022). Understanding followership: A literature review. *Vol. 3: Proceedings of International Conference on Psychological Studies (ICPSYCHE).* https://proceeding.internationaljournallabs.com/index.php/picis/article/view/93

35   Carsten, M. K. (2017). Followership development: A behavioral approach. In M. G. Clark & C. W. Gruber (Eds.), *Leader development deconstructed.* Springer, pp 143–161.

36   Schindler, J. H. (2014). *Followership: What it takes to lead.* Business Expert Press.

37   Kellerman, B. (2008). *Followership: How followers are creating change and changing leaders.* Harvard Business School Press.

38   Mayer, R. C., Davis, J. H., & Schoorman, F. D. (1995). An integrative model of organizational trust. *Academy of Management Review, 20*, 709–734. https://doi.org/10.2307/258792

39   Schoorman, F. D., Mayer, R. C., & Davis, J. H. (2007). An integrative model of organizational trust: Past, present, and future. *Academy of Management Review, 32*, 344–354.

40   Hattie, J., & Timperley, H. (2007). The power of feedback. *Review of Educational Research, 77*, 81–112. https://doi.org/10.3102/00346543 0298487

41   Tuckman, B. W. (1965). Developmental sequence in small groups. *Psychological Bulletin, 63*, 384–399. https://doi.org/10.1037/h0022100

42   Bercovitch, J., & Jackson, R. D. W. (2009). *Conflict resolution in the 21st century: Principles, methods and approaches*. University of Michigan Press.

43   Caspersen, D. (2015). *Changing the conversation: The 17 principles of conflict resolution*. Penguin Books.

44   Ramsbotham, O., Woodhouse, T., & Miall, H. (2011). *Contemporary conflict resolution: The prevention, management and transformation of deadly conflicts* (3rd ed.). Wiley.

Part III

Expanding Horizons in the Age of AI

Chapter 10

Adapting and Growing

The pundits agree that AIs are going to spark major changes in our lives and workplaces. Indeed, current estimates are that by 2034 AIs will either replace or transform about 60% of today's jobs.[1] In fact, as noted in Chapter 1, in some tech fields the "average half-life" of the relevant skills may be less than three years.[2] We are all going to have to adapt quickly and well in order to flourish in the Age of AI.

Adapting requires learning, which is the acquisition of new skills or knowledge. When we adapt, we learn to function in changed circumstances. And learning requires memory. In fact, learning and memory are really just different sides of the same coin: We acquire information via learning and retain it via memory. Without learning, we cannot have a memory of something—and if we don't have a memory of something, at that point we cannot say that we have learned it. When we have successfully learned a skill or some knowledge, we have stored it in our memories.

Many adults seem to regard learning as akin to a session in a dentist's chair, perhaps just a notch or two better than submitting to eye surgery. But learning doesn't have to be painful—or even very effortful. Consider the sort of learning that allows us to summarize a news article that we read earlier in the day, or describe in detail a movie we saw a few days ago, or recount a conversation that took place a week ago. In the vast majority of these cases, we did not try to memorize the event while it was taking place. Even without trying, we picked up and stored such information in our memories. We humans learn an enormous amount through personal experience, observation, and reading—even without intending to learn.[3,4] As we shall see in this chapter, we can draw on the mental processes that underlie such learning to help us adapt.

However, adapting requires more than just learning. In order to adapt, we need three kinds of skills: First, we need to take note when our current practices are no longer working well. Second, we need to chart an alternative path forward. And third, we need to acquire, retain, and apply the

DOI: 10.4324/9781032686653-13

skills and knowledge to follow that new path. We consider each of these topics in turn.

Recognizing Changing Realities

The first step to adapting to change is to identify what has changed and realize that our current practices are no longer serving us well. This initial phase does require effort—we must actively observe and recognize the disparity between what we expect and what actually happens. To do this, we need to exercise metacognition and specific cognitive skills.

Metacognition

Metacognition is cognition about cognition. If you've ever stayed awake for over a day, you may have noticed that you aren't as sharp as usual—that noticing is an example of metacognition. So too is noticing that you can pay attention better after your first cup of morning coffee or tea. Yet another example is when you notice that the way you currently try to learn is not helping much.

Metacognition requires systematic observations. We decide to adapt to changing circumstances only after we we notice that our current practices aren't working well. Because this process does require effort—which we may often try to sidestep or downplay—the Cognitive Amplifier Loop (CAL) can play a special role in helping us learn to employ metacognition more often and more effectively. Consider the following:

Me:

Please ask the user questions that are designed to elicit deeper thought and reflective thinking about their experiences, behaviors, thoughts, and feelings. The goal is to lead the user to engage in metacognition in order to recognize where they need to learn in order to change and adapt. Here are five examples of areas where the user should reflect on the current versus needed state of their skills and knowledge:

1. Personal Growth: "What recent experience made you realize that you need to learn new ways to approach personal growth?" This question encourages the user to identify and reflect on significant experiences that can lead them to realize that they need to change with respect to their personal growth.

2. Professional Development: "What was a challenging situation that you faced at work that did not go as well as it could have, which made you realize that you need to learn new skills or knowledge?" This question should help the user analyze their problem-solving and decision-making skills in a professional context, encouraging them to think critically about shortfalls and what else they may need to learn.

3. Relationships and Interpersonal Skills: "Have you noticed a disconnect between you and someone else, which made you feel awkward and a bit at sea?" Reflecting on this question can enhance understanding of how behaviors and attitudes influence relationships and help the user identify ways to improve relationships.

4. Learning and Education: "What is one concept or skill I learned recently that I have not yet mastered, but have come to realize that it is important and I need to work on it?" This question encourages the user to reflect on their learning process, identify strategies to help them understand complex information, and consider how they can apply these strategies to future learning.

5. Well-Being and Self-Care: "What are three things I need to learn to enhance my well-being and self care?" Learning to enhance well-being and self-care can shift the user's perspective, promote positivity, and encourage them to recognize and appreciate the good in your life, contributing to emotional well-being.

After you ask the question, pause and wait for the user to respond. Do not respond for the user, and wait until they are finished. At that point, observe ways that what they said can help them to learn to engage in reflective thinking more effectively. Then ask whether they want another question. If so, please repeat the process.

AI:

Let's start with the first area: Personal Growth.

Question: What recent experience made you realize that you need to learn new ways to approach personal growth?

Take your time to reflect on this and share your thoughts when you're ready.

Me:

I noticed that I was hesitant to go to a concert that featured only new music, none of which I had heard.

AI:

It sounds like that experience highlighted a hesitation towards new and unfamiliar situations. Reflecting on this, what do you think might be underlying your hesitation? Also, considering this realization, how might you approach similar situations differently in the future to foster personal growth?

Once you've reflected on this, let me know if you would like to move on to the next question or discuss this further.

Me:

Let's discuss this further. I realized that I find listening to new (for me) music difficult and even mildly unpleasant. But if I listen a few times, in many cases I end up appreciating it. How can I learn to remember this realization so I wouldn't automatically dismiss the opportunity to hear new music in the future?

AI:

It's insightful that you've recognized a pattern where repeated exposure leads to appreciation. To help remember and act on this realization in future situations, consider implementing these strategies:

1. Journaling: After attending a concert or listening to new music, write down your initial impressions and how they evolve with more listens. This can help solidify your observations about growing appreciation and make it easier to recall and trust the process.
2. Setting Intentions: Before engaging with new music, set a clear intention to remain open-minded and remind yourself of past experiences where repeated exposure was beneficial. You could

even note down these intentions in a planner or digital note as a pre-event reminder.

3. Mindfulness: Practice mindfulness when encountering new music. Try to notice your immediate reactions without judgment and remind yourself that your perception might change with time. This can help in reducing automatic negative responses.

4. Creating Rituals: Make a ritual of exploring new music regularly, perhaps by setting aside a specific time each week to listen to something outside your usual preferences. This routine can transform anticipation—potentially negative—into curiosity.

5. Social Support: Share your goal of being more open to new music with friends or family. They can encourage you, share their own experiences, or even join you in exploring new genres, which can make the process more enjoyable and memorable.

By actively practicing these strategies, you can reinforce your new understanding and gradually change your automatic responses to new experiences. Would you like to explore another question now, or continue discussing strategies related to this topic?

Me:

This is helpful. Let's stop for now so that I can digest this. Thank you.

AI:

You're welcome! Feel free to return anytime if you want to discuss more or explore other topics. Take care and enjoy your reflections!

Metacognition is complex, and the AI rightfully pointed out that we should use numerous techniques to address it. Indeed, the techniques it recommends have received solid empirical support. For example, James Pennebaker and his colleagues documented the value of journaling for self-insight and coping with troubling emotions and memories.[5] Similarly, numerous studies show that mindfulness meditation can facilitate our ability to engage in metacognition (e.g.,[6]). And part of social support involves seeking feedback, which is especially helpful if it is provided relatively soon after an event and is constructive and non-threatening.[7,8]

Cognitive Skills

We also need to nurture specific cognitive skills to help us recognize when we need to adapt. One such skill is pattern recognition, which operates over both what we perceive and what we conceive. *Perceptual* pattern recognition skills allow us to recognize patterns in what we sense, such as in what we see, hear, or touch. Such processes allow us to recognize a football play, a musical melody, and even to see faces in clouds. *Conceptual* pattern recognition skills allow us to spot patterns in data, events, ideas, and behaviors. Conceptual pattern recognition relies on inductive reasoning, where we draw a generalization from a set of individual cases. For example, if we eat a green apple and discover that it has seeds inside, eat a golden apple and discover seeds inside, and then discover the same thing with another apple, we may inductively reason that apples in general have seeds inside. Such reasoning not only allows us to organize information, but also to predict potential future outcomes. We can pick up on patterns in a host of environmental events that signal the need for change. Such events include advancements in technology (such as AI!), market shifts, and societal changes—as well as consistent shifts in how people are behaving.

Pattern recognition is particularly important when it serves as the grist for our critical thinking mill. As noted in Chapter 3, critical thinking is the ability to analyze and evaluate, which typically leads us to form a judgment or make a decision. Pattern recognition yoked with critical thinking allows us to note when what we are doing is no longer working well enough.

In the present context, we need to decide whether a shift in circumstance or situation is great enough to warrant a personal change. We need to decide whether the ways in which we are falling short are worthy of a concerted response, of learning new skills or knowledge. If we do come to realize that we need to adapt, then we move onto the next phase of the adaptation process, charting a path.

Charting a Path

Once we realize that we need to learn something new, the next step is to decide which specific skills or knowledge we need to learn. In this case, we first need to establish our goal. In many cases, this goal focuses on our job: Do we just want to keep doing our current job, or whatever it has morphed into, or some other job? We can engage in creative problem solving to help us select an appropriate goal (Chapter 4). We can use divergent thinking to cast a broad set of alternatives, and then convergent thinking to choose among them.

Once we have a goal in mind, we can utilize an AI to help us identify the skills and knowledge we need to master in order to achieve that goal. When we are considering a new job, the initial prompt is very simple: We just ask the AI what skills and knowledge are necessary to perform that job. Following this, we can describe our current relevant skills and knowledge to the AI and ask it to assess how difficult it will be for us to acquire the additional skills and knowledge needed for the new job. If the gap is large and our time is short, we may decide to reconsider and engage in another round of the CAL. In this case, we would seek to identify alternative goals that are a better fit to our current skills and knowledge. As usual, engaging the CAL is an iterative process.

Once we decide what particular skills or knowledge we need to learn, we need to identify potential obstacles that stand in the way of learning them. For example, perhaps we need to work full-time to keep bread on the table, and so cannot devote a lot of time to learning. Or perhaps we had difficulty learning certain skills in the past, and need to consider new ways to approach them.

We then need to devise a plan that includes ways either to remove the roadblocks or to sidestep them. This process draws on the sorts of creative problem-solving skills we discussed in Chapter 4. In addition, we would be wise also to formulate a backup plan—just in case.

The CAL can help us at every step of the way. For example, if we are stuck in creating a new goal, the CAL can help us in the divergent thinking process to formulate many different candidates. And it can also help us then achieve clarity when we move into the convergent thinking phase and need to select among the candidates. The same sorts of prompts we considered in Chapter 4 can easily be modified to serve these ends.

Acquiring New Knowledge and Skills

The first two phases—recognizing the need to learn and charting a path forward—are essential for adapting to change, but they are only the prelude. These activities set us up to do the work of actually learning new skills and knowledge. The core of adaptation is learning.

We acquire a lot of what we learn by either reading it or observing it. If we are learning skills, we also need to actually perform the activity, such as driving a car, giving a speech, or singing a song. But even here, we learn a lot before we ever start to perform the skill. Think of what the roads would be like if we couldn't learn a lot about how to drive a car before we ever got behind the wheel. Fortunately, not all learning is via trial-and-error. We can read about the rules of the road and can observe how others drive, and we learn a huge amount by doing so.

We adapt by engaging in two sorts of learning. *Incidental learning* occurs as a byproduct of mentally processing information—such as by carefully observing what someone else does. In this case, we don't need to put any effort into trying to learn. In contrast, *intentional learning* occurs when we do make an effort to acquire skills or knowledge. When we encounter a lot of new material, especially if it is complex, we may need to work in order to master it. We face this challenge in many circumstances, such as when we study for an exam, prepare for a meeting, absorb the contents of a dossier, or learn a large set of new rules and regulations. Intentional learning is a fact of life, and virtually all of us must engage in it from time to time. Thus, it's worth pausing to consider how to do this effectively. We can absorb an enormous amount of information relatively easily if we exploit two fundamental principles (previously referred to as "Maxims"[9]).

Principle 1: Pay Attention and Think It Through

First, *pay attention to what you want to learn and think through what it means*. An enormous amount of learning is a byproduct of simply thinking about something. In general, if we deeply grasp its meaning, we will probably remember it—even if we don't try to do so.[10] This principle explains why we can recall a gripping conversation days later, or summarize a movie that we immersed ourselves in.[11] Although this principle underlies a lot of incidental learning, it can also contribute to intentional learning. Here are some effective techniques we can employ to induce ourselves to pay attention and think deeply about material we want to learn:

Distill. Identify the most important points and note how they are related. To help us do this, we can write bullet points or create a diagram or mind map. For example, after reading a chapter on photosynthesis in a biology textbook, we might write the following bullet points:

- Chlorophyll is green and absorbs sunlight.
- Water is split into oxygen and hydrogen.
- CO^2 is converted into glucose.

As an example in a different sphere, suppose that our romantic partner says that we're "too emotional" and we want to remember the behaviors that produce that impression. The bullet points might look like this:

- I appear to get upset over the small things.
- I seem to get annoyed out of proportion to what's actually happening.
- I keep asking whether something is wrong.

Question. Formulate questions about the material. To do so, we draw on the sorts of critical thinking reviewed in Chapter 3, asking about the source, the logic, alternative interpretations, and so on. For example, after reading an article on AI ethics, we might ask ourselves "What biases exist in AI training data?" "How will those biases affect an AI's responses?" "What arguments are the authors using to support their claims about AI ethics?" "Do these arguments follow logically?" "How else can we look at AI ethics?"

Retrieve. Recall information. The mere act of recalling can strengthen memory. This is related to the "Testing Effect," which occurs when people learn after they take tests—even if they don't receive feedback about their accuracy.[12] For example, after encountering new words in a foreign language, we can make a point of recalling them every few days thereafter, which will cement them in our memory.

Teach. Prepare to explain the material to someone else. To teach, we need to decide what should be put in the foreground and what should be in the background. For example, to teach about rhythm in music, we might foreground the concepts of a "beat," syncopation, accents, and the divisions of notes and rests. In this case, because we were interested in rhythm, we might put the pitch of notes into the background, perhaps using only a single pitch throughout (e.g.,[13]).

Make Relevant. Specify how new information bears on a particular situation of interest. For example, if we are learning about "smart cities," we can iterate through how each innovation—such as in transportation and food production—would affect our own daily lives.

Apply. Think of ways to use new skills or knowledge. For example, when learning about how to write prompts for an AI, we can consider how we could use particular prompts at work.

In all of these cases, we can induce ourselves to pay attention and think through the material, which will help us make sense of it, acquire it, and retain it.

Principle 2: Seek Connections

Now consider a second fundamental principle in the science of learning: *Seek connections between new material and with what you already know.* We should organize material into easily apprehended groups and should find ways to relate such material to what we've learned previously.

By finding such connections, we can integrate new material into our memories, which not only helps that material to "stick," but also makes it easy to locate and dig out later.[14] The following techniques can help accomplish this:

Associate. Associate new information with something we already know. Many memory aids ("mnemonics") rely on creating such associations.[15] For example, if we are learning the name of a new person, we can recall a person we already know who has the same name and visualize the two people shaking hands. Creating this association helps us remember the new person's name.

Analogize. Find an analogy between something new and something we already know. For example, if we are studying electrical circuits, we might relate them to water flowing through pipes.

Compare and Contrast. Identify ways that new information is and is not like something that we already know. For example, if we were learning about transactional leadership and were already familiar with transformational leadership, we could think of how the two are alike and different.

Personalize. Relate new material to something we personally care about. For example, if we are learning about historical events, we can relate them to our own family's history.

Reframe. Fit new information into a familiar framework. For example, we can understand many theories of how nations interact within the same frameworks we rely on to understand how family members interact, considering factors such as what behaviors have been rewarded or punished in the past.

Contextualize. Consider multiple contexts in which the information is relevant. For example, if we are learning about the principle of supply and demand in economics, we can relate this to online dating, prices of lemonade at roadside stands in different parts of the country, or the price of concert tickets.

Learning Styles?

We have focused on techniques that help all humans to learn effectively. But is this the right approach? Perhaps different people have different "learning styles," such as verbal vs. visual. And if we teach Sally or John in ways that fit their individual styles, perhaps they will learn better than if we do not respect these differences. A technique that's good for Sally may not be so good for John, and vice versa. This idea is referred to as the "meshing hypothesis," which states that people learn better when material is presented in a way that is compatible with—meshes with—their own personal learning style.

This meshing hypothesis is intuitively appealing, but the research has not supported it. Indeed, there is no convincing evidence that learning styles actually exist.[16,17,18]

However, there is a wrinkle that makes me take pause before dismissing the learning styles idea completely out of hand. People do have *preferences* for different ways of learning. We like, enjoy, and prefer to engage with material that is presented in a certain way. Even if it is true, as the results from studies indicate, that all of us learn in basically the same way, that we don't in fact have distinct "learning styles," we nevertheless may be more *motivated* to learn if the format appeals to us. We may be more likely to engage in the first place and continue to be engaged if material is presented in a preferred way. My literature search turned up no studies that directly addressed this possibility in real-world contexts.

Such considerations lead us to address motivation, and to consider what motivates us to learn.

Motivating Ourselves to Learn and Adapt

One of our most important metacognitive skills allows us to keep ourselves engaged and not procrastinate—which is essential for both incidental and intentional learning. We need to learn how to keep ourselves motivated to adapt and grow. To do so, we can utilize techniques that grow out of major theories of human motivation. Let's consider the most important theories.

Theories of Motivation

Table 10.1 provides a summary of major theories of motivation, namely Self-Determination Theory (SDT),[19,20,21] Expectancy Theory,[22,23,24] Instinct Theory[25] along with its "Evolutionary Psychology" offshoots,[26,27,28] Maslow's Theory,[29,30,31] and the ARCS Model.[32,33]

Knowing about theories of motivation, and the factors each has identified, can help us set up situations that motivate us. The first step is to note which factors are personally motivating for each of us. We can use the CAL to help. For example, I initially submitted a version of the following prompt where I asked GPT-4 only to draw on Table 10.1. The result struck me as too superficial. So I went to Google and searched for articles that reviewed the scientific literature on motivation, and found quite a few (e.g.,[34,35,36,37]). I downloaded the review by Valarmathie Gopalan and colleagues, and

Table 10.1 Major Theories of Motivation.

Theory	Summary of Theory	Key Criticisms
Self-Determination Theory (SDT)	SDT focuses on intrinsic motivation—motivation driven by inherent satisfaction rather than external rewards. It identifies three intrinsic needs: autonomy, competence, and social relatedness, which are considered essential for psychological health and well-being. SDT is a framework that includes sub-theories, including ones that address the relationships between intrinsic motivation and extrinsic, reward-driven motivation.	Critics argue that SDT might oversimplify the interaction between intrinsic and extrinsic motivations and does not sufficiently address how these motivations are influenced by broader cultural contexts.
Expectancy Theory	This theory explains motivation as a process whereby individuals behave in ways that they expect to lead to particular outcomes. The theory is based on three components: expectancy (belief that effort leads to performance), instrumentality (belief that performance leads to a particular outcome), and valence (value associated with the outcome).	Criticisms focus on the theory's sometimes insufficient consideration of emotional and unconscious influences on the decisions that lead to specific behaviors. Critics also note that expectancy calculations can be highly subjective and influenced by individual perceptions.
Instinct Theory/ Evolutionary Psychology	This theory posits that human behaviors are primarily driven by instincts, which are the result of evolution by natural selection. These instincts include basic survival needs and social behaviors that enhance "genetic fitness" (i.e., success at reproducing). According to this theory, we are born to be inclined to perceive, think, and behave in ways that were advantageous during the evolutionary history of our species.	The main criticism of this theory is that it may oversimplify human behavior, attributing complex actions to basic instincts without considering the role of social, environmental, and cognitive factors. It also may not account for the variability in human behavior across different cultures and contexts.

Table 10.1 (Continued)

Theory	Summary of Theory	Key Criticisms
Maslow's Hierarchy of Needs	This theory posits a hierarchy that orders needs from basic physiological needs to higher-level psychological needs, such as self-actualization. The theory posits that lower-level needs must be satisfied before higher-level needs will influence behavior.	Critics argue that Maslow's theory is overly sequential and does not accurately reflect the complexity of human motivations. Research suggests that needs do not necessarily follow the strict hierarchy proposed, and people may pursue multiple needs simultaneously.
ARCS (Attention, Relevance, Confidence, Satisfaction) Model	Designed for educational settings, the ARCS Model focuses on four elements to motivate learners: Attention, Relevance, Confidence, and Satisfaction. It suggests that learning is motivated by engaging learners' interest, demonstrating relevance, building confidence through scaffolded tasks, and ensuring satisfaction through achievements.	Although the ARCS Model is highly regarded for its practical application in education, it has been criticized for not fully integrating the broader social and emotional contexts that can affect learning, such as cultural backgrounds or previous educational experiences.

attached it to the following prompt when I submitted it. I obtained the following results, which I've truncated after presenting enough to convey the idea:

Me:

Your goal is to figure out what factors especially motivate the user. Attached is a review of theories of motivation, which summarizes a variety of factors that motivate humans. Please construct scenarios that hinge on one factor at a time, and ask the user to indicate how motivated they would be if they were in that situation: 0: Not Motivated, 1: Barely Motivated; 2: Somewhat Motivated; 3: Strongly Motivated. Present the scenarios one at a time, and wait for the user to rate their level of motivation according to this scale. Wait for the user to respond before continuing, and do not respond for them. Continue to present scenarios until you have covered all of the factors in the below, and then tell the user your analysis of which factors they find most motivating. Thank you. [ATTACH REVIEW PAPER]

AI:

Based on the provided motivation theories, I will construct scenarios representing each of the key motivational factors identified in the document. After each scenario, you will rate your level of motivation on a scale from 0 to 3, where 0 is Not Motivated, 1 is Barely Motivated, 2 is Somewhat Motivated, and 3 is Strongly Motivated. We will go through these one at a time.

Intrinsic Motivation Scenario (Challenge): Imagine you are a student who has been given a project that aligns with your interests and is challenging yet achievable. You find the subject fascinating and engaging, and you believe completing this project will give you a deep sense of personal satisfaction. How motivated would you be in this scenario?

Please rate your motivation level, and after your response, we will proceed to the next scenario.

Me:

3

AI:

Extrinsic Motivation Scenario (Recognition): Now envision that you are working on a group project at work. The outcome of this project will be showcased in a company-wide meeting, and significant recognition, along with a potential bonus, will be awarded to the team that delivers the most innovative solution. Your motivation stems from the desire for this external recognition and reward. How motivated would you be in this scenario?

Please rate your motivation level, and after your response, we will proceed to the next scenario.

Me:

2

AI:

Self-Determination Theory Scenario (Autonomy): Imagine you are participating in a professional development program where you have the freedom to choose your learning path and projects based on your personal interests and career goals. This autonomy allows you to engage with content that resonates with you and directly benefits your professional growth. How motivated would you be in this scenario?

Please rate your motivation level, and after your response, we will proceed to the next scenario.

Me:

3

AI:

ARCS Model Scenario (Relevance): Consider you are enrolled in a training program designed to enhance your job performance. The curriculum is tailored to align with your current role and future career aspirations, making the learning content highly relevant and applicable to your daily tasks. This connection between the learning material and your professional life aims to increase your engagement and motivation to learn. How motivated would you be in this scenario?

Please rate your motivation level, and after your response, we will continue with the next scenario.

Me:

3

[TRUNCATED]

AI:

Based on your responses to the various scenarios, it appears that you are highly motivated by situations that provide autonomy, relevance, and a clear effort-performance-reward relationship, as you rated these scenarios with a 3 (Strongly Motivated). You find moderate motivation in scenarios that involve recognition and observational learning, as indicated by your rating of 2 (Somewhat Motivated).

This suggests that factors most motivating to you are those that allow for self-direction and personal relevance, as well as those that offer clear incentives for your effort and performance. Scenarios that involve learning from others or gaining external validation are somewhat motivating but not as influential on your motivation levels.

Overall, it seems you are most driven by intrinsic factors related to autonomy and relevance, as well as extrinsic factors that provide clear rewards for your efforts. These insights can be particularly useful in tailoring your work and learning environments to maximize your motivation and engagement.

We can evaluate the AI's judgments by using a technique we discussed in Chapter 5. Namely, we can access our subconscious preferences by having the AI generate an assessment and then we consider how we feel about it. We can determine whether the AI's judgments ring true. This process can lead to personal insights about what sorts of factors motivate us. This sort of personalized assessment is difficult and expensive to achieve without an AI.

Once we have insight into what motivates us, we can either select situations that include those factors or work to create them. For example, if autonomy is important to us, we can ask to be assigned to teams that are given a general mandate, but are not micromanaged from above. Moreover, if we know that such situations are highly motivating for us, we can advocate for them, making reasoned arguments about why such teams are more nimble and likely to do something new.[38] Indeed, the CAL can help us devise and make such arguments.

In general, an AI can help us learn and adapt by providing scenario-based exercises, immediate feedback, and focused practice on what we find most difficult. It can also help us to manage working memory load,[39,40] and can provide a huge range of types of active learning.[41] In all cases, we can use the CAL to induce the AI to give us material that is not only at the right level for us, given our relevant background, but also incorporates examples and illustrations that we find engaging.

Flourishing with the CAL

Learning new skills and knowledge clearly can lead to a sense of autonomy and control because learning gives us tools to navigate our environments. New skills and knowledge can even guide us in finding a good match when looking for a new job—or looking for a new assignment at our existing job. Improving our ability to learn can enhance our job satisfaction and can allow us to move up in an organization, which in turn provides greater financial security. In addition, we noted earlier that to flourish we need to grow as people, to develop our abilities and talents. Learning is a key to achieving these goals.

All our adapting and growing takes place against the backdrop of our broad life goals, explicit or implicit. These goals are what propel us forward, and what ultimately underlie our sense of meaning and purpose. To flourish, we need to formulate such goals and make progress in obtaining them. The final chapter of this book addresses ways we can adapt our life goals in the Age of AI, and enlist AIs to help us achieve these goals.

Digging Deeper

As usual, we end with suggestions of videos that provide an accessible, more in-depth treatment of the material discussed in this chapter, and also provide search terms that can help to locate new videos on the topic.

Metacognition: The Skill that Promotes Advanced Learning: https://www.youtube.com/watch?v=elZFL4FLVLE
Introduction to Inductive and Deductive Reasoning (Includes Activity): https://www.youtube.com/watch?v=G4icPtEY8f8

- "Metacognition and personal growth"
- "Journaling for self-reflection and learning"
- "Meditation and mindfulness for metacognition"
- "Seeking feedback for personal development"
- "Critical thinking exercises for self-improvement"
- "Pattern recognition in decision making"
- "Inductive reasoning and adapting to change"
- "Adapting to changing professional landscapes"
- "Lifelong learning in the Age of AI"

How to Achieve Your Most Ambitious Goals | Stephen Duneier: https://www.youtube.com/watch?v=TQMbvJNRpLE

The Science of Learning: How to Turn Information into Intelligence | Barbara Oakley | Big Think: https://www.youtube.com/watch?v=1FvYJhpNvHY

Meet a Game Changer: Art Markman: https://www.youtube.com/watch?v=ep15Om07pyA

- "Effective learning techniques for adults"
- "Connecting new information to existing knowledge"
- "Retrieval practice and learning"
- "Distributed practice and spaced repetition"
- "Elaborative interrogation and self-explanation"
- "Interleaved practice and learning"
- "Using AI to support learning and skill acquisition"
- "Personalized learning with AI"
- "Debunking the myth of learning styles"
- "Learning strategies vs. learning styles"
- "Motivation and engagement in learning"
- "Adapting to change through continuous learning"

The puzzle of motivation | Dan Pink | TED: https://www.youtube.com/watch?v=rrkrvAUbU9Y

ARCS: A Conversation with John Keller: https://www.youtube.com/watch?v=E1ugbX2EKN0

- "Self-Determination Theory and motivation"
- "Intrinsic vs. extrinsic motivation"
- "Autonomy, competence, and relatedness in motivation"
- "Expectancy Theory of motivation"
- "Instinct Theory and evolutionary psychology"
- "Maslow's Hierarchy of Needs"
- "ARCS Model of Motivation in education"
- "Leveraging AI to identify personal motivational factors"
- "Scenario-based learning and motivation"
- "Immediate feedback and motivation"
- "Managing cognitive load in learning"
- "Active learning strategies and motivation"

References

1   Georgieva, K. (2024, January 14). AI will transform the global economy. Let's make sure it benefits humanity. *IMF blog*. www.imf.org/en/Blogs/Articles/2024/01/14/ai-will-transform-the-global-economy-lets-make-sure-it-benefits-humanity

2   Tamayo, J., Doumi, L., Goel, S., Kovacs-Ondrejkovic, O., & Sadun, R. (2023, September–October). Reskilling in the Age of AI. *Harvard Business Review*. https://hbr.org/2023/09/reskilling-in-the-age-of-ai

3   Brown, P. C., Roediger, H. L. III, & McDaniel, M. A. (2014). *Make it stick: The science of successful learning*. Belknap Press/Harvard University Press.

4   Kosslyn, S. M. (2023). *Active learning with AI: A practical guide*. Alinea Learning.

5   Pennebaker, J. W. (2004). *Writing to heal: A guided journal for recovering from trauma & emotional upheaval*. New Harbinger Publications; Distributed in Canada by Raincoast Books.

6   Creswell, J. D. (2017). Mindfulness interventions. *Annual Review of Psychology, 68*, 491–516. https://doi.org/10.1146/annurev-psych-042716-051139.

7   Hattie, J., & Clarke, S. (2018). *Visible learning: Feedback*. Routledge. doi:10.4324/9780429485480.

8   Wiśniewski, B., Zierer, K., & Hattie, J. (2020). The power of feedback revisited: A meta-analysis of educational feedback research. *Frontiers in Psychology: Section on Educational Psychology, 10*. www.frontiersin.org/journals/psychology/articles/10.3389/fpsyg.2019.03087/full

9   Kosslyn, S. M. (2017). The science of learning. In S. M. Kosslyn & B. Nelson (Eds.), *Building the intentional university: Minerva and the future of higher education*. MIT Press, pp. 149–164.

10 Willingham, D. T. (2023). *Outsmart your brain: Why learning is hard and how you can make it easy*. Gallery Books.

11 Kosslyn, S. M. (2020). *Active learning online: Five principles that make online courses come alive*. Alinea Learning.

12 Roediger, H. L., & Karpicke, J. D. (2006). Test-enhanced learning: Taking memory tests improves long-term retention. *Psychological Science, 17*, 249–255. https://doi.org/10.1111/j.1467-9280.2006.01693.x

13 Bellson, L. (1985). *Modern reading text in 4/4 for all instruments*. Alfred Music.

14 Kosslyn, S. M. (2023). *Active learning with AI: A practical guide*. Alinea Learning.

15 Foer, J. (2011). *Moonwalking with Einstein: The art and science of remembering everything*. Penguin Press.

16 Cuevas J. (2015). Is learning styles-based instruction effective? A comprehensive analysis of recent research on learning styles. *Theory and Research in Education, 13*, 308–333. doi:10.1177/1477878515606621.

17 Pashler, H., McDaniel, M., Rohrer, D., & Bjork, R. (2008). Learning styles: Concepts and evidence. *Psychological Science in the Public Interest, 9*, 105–119. https://doi.org/10.1111/j.1539-6053.2009.01038.x.

18   Willingham, D. T., Hughes, E. M., & Dobolyi, D. G. (2015). The scientific status of learning styles theories. *Teaching of Psychology, 42*, 266–271. https://doi.org/10.1177/0098628315589505.

19   Deci, E. L., & Ryan, R. M. (2012). Motivation, personality, and development within embedded social contexts: An overview of self-determination theory. In R. M. Ryan (Ed.), *Oxford handbook of human motivation*. Oxford University Press, pp. 85–107.

20   Ryan, R. M., & Deci, E. L. (2000). Self-determination theory and the facilitation of intrinsic motivation, social development, and well-being. *American Psychologist, 55*, 68–78. doi:10.1037/0003-066X.55.1.68.

21   Ryan, R. M. & Deci, E. L. (2017). *Self-determination theory: Basic psychological needs in motivation, development, and wellness*. The Guilford Press https://doi.org/10.1521/978.14625/28806.

22   Vroom, V. H. (1964). *Work and motivation*. Wiley.

23   Vroom, V. H. (2005). On the origins of expectancy theory. In K. G. Smith & M. A. Hitt (Eds.), *Great minds in management*. Oxford University Press, pp. 239–258. https://doi.org/10.1093/oso/9780199276813.003.0012

24   Montana, P. J., & Charnov, B. H. (2008). *Management* (4th ed.). Barron's Educational Series.

25   McDougall, W. (1932). *The energies of men: A study of the fundamentals of dynamic psychology*. Methuen & Co.

26   Bernard, L. C., Mills, M., Swenson, L., & Walsh, R. P. (2005). An evolutionary theory of human motivation. *Genetic, Social, and General Psychology Monographs, 131*, 129–84. doi:10.3200/MONO.131.2.129-184.

27   Cosmides, L., & Tooby, J. (2013). Evolutionary psychology: New perspectives on cognition and motivation. *Annual Review of Psychology, 64*, 201–229. doi:10.1146/annurev.psych.121208.131628.

28   Pinker, S. (2007). *The language instinct: How the mind creates language*. Harper Perennial Modern Classics.

29   Maslow, A. (1943). A theory of human motivation. *Psychological Review, 50*, 370–396. https://doi.org/10.1037/h0054346.

30   Tay, L., & Diener, E. (2011). Needs and subjective well-being around the world. *Journal of Personality and Social Psychology, 101*, 354–365. doi:10.1037/a0023779.

31   Wahba, M. A., & Bridwell, L. G. (1976). Maslow reconsidered: A review of research on the need hierarchy theory. *Organizational Behavior and Human Performance, 15*, 212–240. doi:10.1016/0030-5073(76)90038-6.

32   Keller, J. M. (1979). Motivation and instructional design: A theoretical perspective. *Journal of Instructional Development, 2*, 26–34.

33   Keller, J. M. (1987). Development and use of the ARCS model of instructional design. *Journal of Instructional Development, 10*, 2–10.

34   Filgona, J., Sakiyo, J., Gwany, D. M., & Okoronka, A. U. (2020). Motivation in learning. *Asian Journal of Education and Social Studies, 10*, 16–37. https://doi.org/10.9734/ajess/2020/v10i43027

35   Gopalan, V., Bakar, J. A. A., Zulkifli, A. N., Alwi, A., & Mat, R. C. (2017). A review of the motivation theories of learning. *AIP Conference Proceedings*, 1891, 020043. https://doi.org/10.1063/1.5005376

36   Graham, S., & Weiner, B. (2012). Motivation: Past, present, and future. In K. R. Harris, S. Graham, T. Urdan, C. B. McCormick, G. M. Sinatra, & J. Sweller (Eds.), *APA educational psychology handbook, Vol. 1: Theories, constructs, and critical issues*. American Psychological Association, pp. 367–397. https://doi.org/10.1037/13273-013

37   Lai, E. R. (2011). *Motivation: A literature review*. www.johnnietfeld.com/uploads/2/2/6/0/22606800/motivation_review_final.pdf

38   Christensen, C. M. (2016). *The innovator's dilemma: When new technologies cause great firms to fail*. Harvard Business Review Press.

39   Clark, R. C., Nguyen, F., & Sweller, J. (2005). *Efficiency in learning: Evidence-based guidelines to manage cognitive load*. Wiley.

40   Sweller, J. (1988). Cognitive load during problem solving: Effects on learning. *Cognitive Science, 12*, 257–285. doi:10.1207/s15516709cog1202_4.

41   Kosslyn, S. M. (2023). *Active learning with AI: A practical guide*. Alinea Learning.

Chapter 11

Curating Life Goals

We engage the Cognitive Amplifier Loop (CAL) because we have a problem we need to solve or a question we want to answer. Such goals are often a result of requirements for doing a job or finishing a specific task. These sorts of short-term goals are distinct from the long-term goals that we set for our lives, which tend to be larger, more personal, and change less often than the goals that spur us to engage the CAL. Nevertheless, our life goals—explicit or implicit—affect many aspects of our daily lives, including how and when we engage with AIs.

We all create and update life goals. Although the task faced by a high-school student and a 40-year-old professional might seem very different, they share many common features. In all cases, we need to identify what the possibilities are, what inspires us, what seems feasible for us, and what we can see ourselves doing.

We need to distinguish between two aspects of our life goals: On the one hand, our *human goals* are what drives our personal lives. These goals focus on what kind of a human being we want to be, and are deeply rooted in our values, relationships, and world view. On the other hand, our *professional goals* are what governs much of our day-to-day work lives, and are rooted in the practical skills and knowledge we utilize in our job. The two sorts of goals are intertwined, and together confer purpose and meaning in our lives.

Whether we are creating our first tentative human and professional goals or revising our mature goals to be more satisfying, we need to learn more than just the practical skills and knowledge we use at work. In particular, many of our human goals may be informed by skills and knowledge that at first glance may not seem useful. In the next section we explore this idea.

DOI: 10.4324/9781032686653-14

Roles of the Humanities in Setting Human Goals

The humanities are often considered areas of knowledge that are not useful for lay people. The humanities address subjects such as art, classics, dramatic arts, history, literature, music, and philosophy. Yes, the academics who study the humanities and the practitioners who specialize in it find this knowledge useful—but for the rest of us, what we learn in the humanities may often just sit in the back rooms of our minds, gathering neural dust. But this need not be so: As the similarity in their names may suggest, the connection between "human" goals and the "humanities" may run deep.

Defining the Range of Human Goals

How can knowledge from the humanities inform our human goals? To put the humanities to this use, we may need to adopt a new way of thinking about them. Stay with me for a couple of minutes as we explore an analogy, based on my personal experience: I was surprised when I discovered that my university considered history to be a social science. I realized that I really didn't know much about the formal discipline of history, so I arranged a series of lunches with individual senior faculty members in that department, to talk about what the study of history is all about. I was disoriented by what I learned: First, none of the scholars I talked to subscribed to George Santayana's famous dictum, "Those who cannot remember the past are condemned to repeat it" (p. 284).[1] The faculty members I talked to emphasized that each historical event is unique, a result of a particular combination of circumstances and factors that interact in ways that will never be precisely duplicated. Thus, I was told, learning about the past won't directly apply to the present or future. Moreover, these historians told me that they were not trying to discover general principles that govern how events unfold over time. Memorably, one told me that as far as he was concerned, "The 17th century could be an island off the coast of India" in the present.

I was taken aback by these sorts of assertions, and the various anecdotes and observations that accompanied them. For example, one of the historians pointed out that no historian had predicted the fall of the Berlin Wall—or any other major historical event. I pondered all of this for a few weeks, and came to a hypothesis: It seemed to me that these historians were in fact studying psychology, but not the sort that is typically studied in psychology departments. Having listened to these scholars, I had the strong sense that they were really interested in human nature and how it is shaped by different circumstances, such as those that existed in Nazi Germany vs. the American frontier. These historians seemed to be analyzing how such different large-scale contexts shape what it is to be human.

Admittedly, I consulted a very small sample, which may not represent the field at large. But even when other historians do appeal to the idea of learning from the past to avoid problems in the present, they appeal to facets of human nature.[2] That is, they aren't formulating "Laws of History," but rather are characterizing how we humans behaved in specific previous situations—and then predicting that we will behave similarly in subsequent similar situations.

The idea that at least some historians want to understand how the environment shapes human nature may parallel the concept of "Range of Reaction" in genetics (also sometimes called the "Reaction Norm" or "Norm of Reaction").[3,4] Here's the key idea: Our genes specify the range of a possible characteristic, and the environment sets each person within that range. For example, consider height: people aren't going to be 2 feet tall or 10 feet tall, and each of us has genes that set a particular, narrower range within this span. But where we end up within that range depends on the environment. For instance, Japanese men today are, on average, almost 4 inches taller than were their great-grandfathers. The genes haven't changed; rather, today's Japanese have better nutrition, possibly less overall severe stress, and experience other differences in the environment.

Now let me close the loop and bring this back to how we can think about human goals. First, we've already considered the idea that history in particular can help define the range of ways to be human. The classics, dramatic arts, literature, and philosophy do the same. This is analogous to identifying the upper and lower limits of the Range of Reaction in genetics. By studying these fields, we can learn a lot about the ways each of us can be humans—what kinds of people we can be, with what values and what sorts of worldviews.

Once we have some idea of the space of possible human goals, we can then progress through that space based in part on what we learn from the humanities. In particular, many aspects of the humanities allow us to "stand in the shoes" of others, which allows us to benefit from their experiences. True, we cannot entirely adopt another person's perspective, but we do have certain things in common simply because we are all human beings. For example, we all live in a world where actions have consequences. In the previous chapter we noted that there is no need for us to engage in trial and error learning if we can observe the consequences of someone else's actions—and that method applies not only to direct observation, but also to the kinds of "second hand" observations provided by many works in the humanities. Literature, dramatic arts, history, classics, and philosophy are replete with such "case studies." For example, when we read a novel, we put ourselves in the shoes of the protagonist and the other characters. As we read, we can reflect on what makes us comfortable, what makes us less at ease, what surprises us, what makes us take pause, and so on. We can soak in the experiences as if we lived them. Indeed, we can define a type of "good" literature based on the extent to which it induces such authentic experiences.

The humanities also provide sources of direct and indirect experiences that help us understand ourselves, which is critical for charting our paths forward. As the Greek philosopher Plato quotes his teacher, Socrates, as having said, "The unexamined life is not worth living."[5] Some of the humanities can produce insights about ourselves and our world by putting us in a specific state of mind. Art and music often do this, as does drama. These mental states can help us appreciate how we see the world and how we see ourselves in it. Similarly, we can regard philosophy as a way to examine the foundations of beliefs and ethics that should inform our life goals.

In short, learning the humanities can help us formulate and update our human goals while the landscape shifts beneath our feet as AIs permeate all aspects of society. To draw on the humanities from this perspective, we would probably not learn them in the traditional way, as "knowledge for knowledge's sake." Rather, we would use the material to shape long-term objectives we can target. To do this well, we can consider the material from four levels of analysis, based on those described by Russell Ackoff: [6]

We start by considering the *data* that—explicitly or implicitly—underlie the material. Data are uninterpreted results of qualitative observations or quantitative measurements. They may be conveyed as numbers, characters, words, or images. We need to determine whether the novel, play, or other humanistic work rests on solid observations.

We next should consider how the author of the work interpreted the data. When we interpret or analyze data, it becomes *information*. Although information is based on data, it emerges from interpretations that are provided or generated, such as when an author makes a deduction or reaches a conclusion. Are such interpretations coherent?

Next, we need to put information in context, which produces *knowledge*. To have knowledge, we need to relate information to other things we know. We need to associate it and integrate it. We need to decide whether the knowledge we derive from some aspect of the humanities makes sense in a specific context. This requires bringing to bear much of what we have previously discussed. For example, we need to consider our biases, emotional reactions, and so on.

Finally, when we put knowledge into perspective, we can achieve *wisdom*. Wisdom allows us to know how best to use our knowledge. This sounds very lofty, but we should take a step back and regard what we glean from the humanities from the broadest perspective—and ensure that this can help us achieve our human goals.

To flourish in the Age of AI, we need to foreground the aspects of the humanities that inform our human goals, which ideally would lead to a measure of wisdom. This is not to say that we should examine only some parts of these fields, which might foreclose certain options down the road. Rather, the idea is to highlight the aspects of the humanities that can guide us in our life journey. This approach makes this material come alive, so that it is not simply "inert ideas," as Sir Alfred North Whitehead worried.[7]

Using the CAL to Consider Human Goals

Human goals may sometimes seem vague and abstract, but they play out in how we live our daily lives. What kind of person we want to be often translates to how we want to behave in a specific, human context. To help us think through the space of alternatives for our human goals, where we are now, and where we want to be, we can utilize the following prompt:

Me:

I want you to play the role of a coach and wise counselor. I have several concerns about my personal future. Please ask me to express such a concern. Then wait for me to respond. Do not move on until I've sent you a concern. Then help me to reason about this concern by asking me leading questions. After I've answered each question, help me reason about the trade-offs that are implied. Thank you.

AI:

I'm ready to support you in navigating through your concerns. Please express one of the concerns you have about your personal future, and we'll explore it together.

Me:

What would I do with myself if I retired?

AI:

That's a significant and thoughtful question to consider. Let's explore it together.

How do you currently spend your time outside of work? What hobbies or activities do you enjoy?

Me:

I have two hobbies: The French language (at which I am mediocre) and playing electric bass guitar.

AI:

Those are interesting hobbies! Let's dive a bit deeper into each.

French Language:

- What motivates you to learn French? Is it the love for the culture, the challenge of learning a new language, or something else?
- Have you ever thought about how you could expand this interest in retirement? For example, traveling to French-speaking countries, joining language groups, or even teaching French to beginners?

Playing Electric Bass Guitar:

- What do you enjoy most about playing the bass guitar? Is it performing, the personal satisfaction of mastering a song, or maybe the social aspect of playing with others?
- Have you considered ways to deepen this hobby in retirement, such as joining a band, recording music, or volunteering your skills in community events?

These questions aim to help us understand not just the activities you enjoy, but also what aspects of those activities fulfill you. This understanding can guide us in envisioning a retirement that is both active and personally meaningful.

Me:

About French and Bass Guitar: They share two main motivations: 1) I am not "naturally inclined" to be good at languages or music, and must work to learn them. I find it satisfying to note progress. 2) I like the social pieces, both speaking to people in French and playing music with others. The problem is that I can only do each of these activities for some portion of the day, say half. What would I do with the other half?

AI:

Your motivations reveal a lot about what you might find fulfilling in retirement. It's commendable that you enjoy challenging yourself and appreciate the social aspects of your hobbies. Let's consider how these motivations can help us think about the other half of your day.

Seeking Challenges and Learning:

- Beyond languages and music, are there other areas or skills you've been curious about but haven't had the time to explore? This could be anything from arts and crafts, to technology, to physical fitness or even cooking.
- How might you structure learning and challenges in these new areas to give you a similar sense of progression and accomplishment?

Expanding Social Engagement:

- Given your enjoyment of social interactions, have you considered volunteer work, clubs, or organizations where you could meet new people and contribute meaningfully?
- How about teaching or mentoring in your areas of expertise? Sharing your journey in mastering French and the bass guitar could inspire and help others, offering you new social connections and a sense of purpose.

Integration and Balance:

- Considering the balance of your day, how important is it for you to have a varied schedule versus deep diving into a few select activities?
- Could some of your interests naturally extend into other social or learning opportunities? For instance, might learning French cooking or the history of music offer new avenues of exploration that intersect with your current hobbies?

It's also valuable to think about how some of these activities might not just fill your day but could overlap in enriching ways, enhancing both the challenge and the social aspects you enjoy. What are your thoughts on these suggestions?

Me:

Thanks. You've given me a lot to think about.

AI:

You're welcome! I'm glad to hear that our conversation has given you some avenues to explore. Remember, retirement can be a deeply rewarding phase of life, full of opportunities to grow, connect, and enjoy the fruits of your labors in new and exciting ways. If you have any more questions or need further guidance down the line, feel free to reach out. Best wishes on your journey!

Again, in this case the main value of the AI is to get us to think deeply, noticing what rings true and what feels "off." As we've seen in previous chapters, an AI can help us think more clearly and expansively, but it won't substitute for our own reactions and conclusions.

Charting Professional Goals

Part of our life goals must include our professional goals. Our jobs are a large part of our lives. An initial question we should ask is, "Which current jobs will survive, which will disappear, and which will change fundamentally to accommodate new realities in the Age of AI?" Although nobody can answer with certainty, we can get reasonably far by considering the implications of the idea that humans are better than current AIs at responding in open-ended situations that require taking context into account. To get a sense of what sorts of current jobs are likely to survive, I asked GPT-4 the following, and present its response right below that prompt:

Me:

AI is going to replace many jobs and affect many others. Let's posit that the jobs most likely to survive require managing open-ended situations that require taking context into account. Please give me a list of jobs that fall into three categories: 1) Those that are likely to be replaced by AI; 2) Those that are likely to require working with AI but not be replaced by it; and 3) Those that will barely be touched by AI, if at all. Please give me 10 jobs in each category, and provide a brief justification for why you put each job in that particular category.

AI:

Here's a list of jobs categorized based on their potential interaction with AI:

1) **Jobs likely to be replaced by AI:**

   - **Telemarketing**: Highly scripted and repetitive, making it suitable for automation.
   - **Data Entry Clerks**: Involves routine, structured tasks easily handled by software.
   - **Payroll Managers**: Processes can be automated through sophisticated algorithms.
   - **Travel Agents**: Online platforms can provide personalized travel solutions efficiently.
   - **Assembly Line Workers**: Robots can perform repetitive, manual tasks more efficiently and without fatigue.
   - **Bank Tellers**: Automated teller machines and online banking systems can handle routine transactions.
   - **Bookkeeping Clerks**: Software can accurately manage data entry and financial records.
   - **Proofreaders**: AI can detect grammatical errors and inconsistencies quickly.
   - **Retail Cashiers**: Automated checkout systems and self-service kiosks are becoming more common.
   - **Market Research Analysts**: AI can process large datasets to predict market trends.

2) **Jobs likely to require working with AI but not be replaced by it:**

   - **Doctors**: AI can assist with diagnostics and patient monitoring, but human oversight is crucial.
   - **Lawyers**: AI can help with research and data analysis, but complex legal reasoning requires human judgment.
   - **Marketing Managers**: AI can analyze data and automate routine tasks, but strategic decision-making will need a human touch.

- **HR Managers**: AI can streamline recruitment and administrative tasks, but human interaction is essential for handling sensitive issues.
- **Journalists**: AI can help sift through data and generate initial reports, but investigative journalism requires deep human insight.
- **Software Developers**: AI can optimize code and perform testing, but creative software solutions require human ingenuity.
- **Teachers**: AI can personalize learning and automate grading, but human interaction is vital for motivating and mentoring students.
- **Urban Planners**: AI can analyze data for planning, but community engagement and policy considerations need human skills.
- **Customer Service Managers**: AI can handle routine inquiries, but complex customer issues require human empathy.
- **Financial Analysts**: AI can process large volumes of financial data, but making high-level financial decisions involves nuanced judgment.

3) **Jobs that will barely be touched by AI, if at all:**

- **Psychotherapists**: Emotional intelligence and understanding human feelings are beyond AI's capabilities.
- **Social Workers**: Requires empathy, cultural understanding, and personal interaction that AI cannot replicate.
- **Clergy**: Spiritual leadership and personalized religious guidance are deeply human roles.
- **Chief Executives**: Strategic leadership and corporate decision-making involve complexities that AI can't handle.
- **Human Resources Executives**: Dealing with organizational culture and employee relations requires nuanced human interactions.
- **Event Planners**: Creativity and adaptability in dynamic environments are key aspects.
- **Politicians**: Involves negotiation, persuasion, and a deep understanding of human desires and societal norms.

- **Artists**: Creativity and expression in art forms are inherently human traits.
- **Scientists (especially in fields requiring conceptual and complex problem-solving)**: Deep theoretical thinking and innovative experiments need a human approach.
- **Ethicists**: Moral reasoning and ethical decision-making involve complex human values.

These categorizations consider current trends in AI development and its application in various fields, highlighting the balance between tasks that AI can automate and areas where human skills are irreplaceable.

This list spans a large range of options, and new types of jobs will undoubtedly materialize as time goes on. It's notable that so many of the jobs that are likely to survive will probably be fulfilling and provide personal satisfaction. However, it's sobering to note that the AI may have been too conservative: AIs are already doing passable work in at least some of the jobs it expected to be barely touched by AI, such as psychotherapy.

As each of us looks ahead to chart our professional aspirations, we initially may form ambiguous, abstract, or vague goals. We can make them more concrete and tractable by breaking them down into a set of more specific, concrete goals. For example, say we want to become a manager in a human resources (HR) department, which is one of the jobs noted above that is likely to survive. We can use the CAL to help firm up our plans. Unlike conducting a Google search or the like, the AI can personalize its responses, tailoring them to our own particular needs and preferences. For instance, let's say we are a marketing specialist aiming for a management position in HR. We could ask the AI the following (this was to Claude 3 Opus):

Me:

I am a sales professional who wants to move into a management career in human resources. I've been working in retail clothing stores for over 10 years, and enjoy interacting with people. I like the idea of helping people function well within organizations. What sorts of skills and knowledge do I need to acquire in order to qualify for such a job?

AI:

To transition from a sales role to a management career in human resources (HR), you'll need to develop a diverse set of skills and acquire specific knowledge. Here are some key areas to focus on:

1. HR Fundamentals: Gain a solid understanding of HR principles, practices, and terminology. This includes topics like recruitment, employee relations, performance management, compensation and benefits, and organizational development.
2. Employment Laws and Regulations: Familiarize yourself with relevant labor laws, such as the Fair Labor Standards Act (FLSA), Equal Employment Opportunity (EEO) laws, and the Family and Medical Leave Act (FMLA).
3. People Management Skills: Develop skills in leadership, communication, conflict resolution, and coaching. These skills are essential for managing teams and fostering a positive work environment.
4. Analytical and Problem-Solving Skills: HR managers often deal with complex issues that require data analysis and problem-solving abilities. Strengthen your skills in these areas to make informed decisions.
5. Technology Proficiency: Become proficient in using HR software, such as Human Resource Information Systems (HRIS) and Applicant Tracking Systems (ATS).
6. Industry Knowledge: Stay updated on HR trends, best practices, and challenges specific to your target industry.

To acquire these skills and knowledge, consider the following steps:

1. Education: Pursue a degree or certification in HR, such as a Bachelor's in Human Resources Management or a Professional in Human Resources (PHR) certification.
2. Volunteer or Intern: Look for opportunities to gain hands-on experience in HR through volunteering or internships.
3. Network: Attend HR events, join professional associations, and connect with HR professionals to learn more about the field and potential job opportunities.

4. Self-Study: Read HR-related books, articles, and online resources to stay informed about the latest trends and best practices.
5. Transferable Skills: Highlight your transferable skills from sales, such as communication, interpersonal skills, and the ability to work in a fast-paced environment.

By combining your existing skills with targeted education and experience in HR, you can position yourself as a strong candidate for HR management roles.

Notice how most of the competencies the AI flagged draw on the material we've considered in this and previous chapters. The exceptions typically are skills and knowledge that bear on a specific job. But what about coaching and networking skills? We haven't considered them previously. In a second iteration of the loop, we can ask the AI to drill down on one of these goals. For example, within the same thread, I asked the AI:

Me:

How should I acquire the skills that involve networking?

AI:

Networking is a crucial skill for advancing your career in HR. Here are some strategies to help you acquire and develop networking skills:

1. Join Professional Associations: Become a member of HR professional associations, such as the Society for Human Resource Management (SHRM) or the National Human Resources Association (NHRA). These organizations offer networking events, conferences, and online forums where you can connect with other HR professionals.
2. Attend Industry Events: Participate in HR conferences, seminars, workshops, and trade shows. These events provide excellent

opportunities to meet people, learn about the latest trends, and exchange ideas with experts in the field.

3. Leverage Online Platforms: Use professional networking platforms like LinkedIn to connect with HR professionals, join relevant groups, and participate in online discussions. Engage with content shared by others and share your own insights to establish your presence.

4. Reach Out to Professionals: Identify HR professionals in your target companies or industries and reach out to them for informational interviews. Ask about their career paths, challenges, and advice for someone looking to transition into HR.

5. Volunteer: Offer your skills and time to local business associations, non-profits, or community organizations. Volunteering allows you to expand your network while contributing to a cause you care about.

6. Seek Mentorship: Find a mentor in the HR field who can guide you, provide advice, and introduce you to their professional connections. Mentors can be found through professional associations, alumni networks, or online platforms.

7. Follow Up and Nurture Relationships: After meeting new contacts, follow up with a personalized message or email. Keep in touch regularly by sharing relevant articles, offering assistance, or meeting for coffee to maintain and strengthen your relationships.

Remember, networking is about building genuine relationships based on mutual interests and value exchange. Focus on how you can contribute to others while also learning from their experiences. Consistency and authenticity are key to successful networking.

Most of these suggestions are about how to put ourselves in situations where we can engage in networking, not about actually acquiring the skills to network successfully. This is a good example of using the CAL: I needed to iterate, engaging in another cycle by updating the prompt. I did so below, in the same thread.

Me:

But what should I do to actually engage in networking successfully?

AI:

To engage in networking successfully, you need to be proactive, genuine, and strategic in your approach. Here are some practical tips to help you network effectively:

1. Prepare an Elevator Pitch: Develop a concise, compelling introduction that summarizes your background, skills, and career goals. Practice delivering it confidently and naturally.
2. Set Networking Goals: Define specific goals for each networking opportunity, such as learning about a particular company, seeking advice on transitioning to HR, or connecting with a certain number of new contacts.
3. Research Attendees: Before attending events, research the attendees and identify the people you'd like to meet. This helps you prioritize your time and tailor your conversations.
4. Ask Questions and Listen Actively: When engaging with others, ask open-ended questions about their experiences, challenges, and insights. Listen attentively and show genuine interest in their perspectives.
5. Offer Value: Share your own knowledge, experiences, or resources that could be helpful to your contacts. This could include industry articles, job leads, or connections to others in your network.
6. Be Authentic and Confident: Be yourself and communicate confidently about your skills, experiences, and aspirations. Authenticity builds trust and helps you form genuine connections.
7. Follow Up: After meeting someone, send a personalized follow-up message within 24–48 hours. Mention something specific from your conversation and express your appreciation for their time. If you committed to sending resources or making an introduction, do so in a timely manner.
8. Nurture Relationships: Keep in touch with your contacts regularly, even when you don't need immediate assistance. Share updates, congratulate them on their successes, and offer ongoing value to maintain strong relationships.

9. Seek Opportunities to Help: Look for ways to assist your contacts, such as making introductions, sharing job opportunities, or providing recommendations. Helping others builds goodwill and increases the likelihood that they'll reciprocate in the future.

10. Evaluate and Adjust: Reflect on your networking experiences and assess what's working well and what could be improved. Continuously refine your approach based on feedback and results.

Remember, networking is a skill that improves with practice. The more consistently and strategically you engage in networking activities, the more comfortable and effective you'll become in building meaningful professional relationships.

I found this response helpful. I could have continued by asking the AI to create a role-playing game to help me practice those skills. This again emphasizes the iterative nature of the CAL. We human users need to be actively engaged, keeping the reason for engaging the AI in mind and evaluating the results accordingly.

We can employ the CAL to help us conduct a thorough feasibility analysis of career plans, which can ensure that our ambitions are not just dreams but rather are grounded in the reality of the current landscape. Human and professional growth must take place in a surrounding societal context. And AIs are no doubt going to have a massive impact on that context. This is already evident, even with the current levels of AI. Given the highly competitive climate among AI companies, we can be sure that these systems will only become increasingly powerful and sophisticated.

Flourishing with the CAL

In this chapter we focused on a fundamental requirement to truly flourish in the Age of AI: Formulating our life goals and pursuing them in ways that give meaning and purpose to our lives. We've seen that even in this most human of endeavors, an AI can help.

In the previous pages I've argued that we humans need to double down on mastering the skills and knowledge that AIs will not easily master in the near future. I've claimed that, by virtue of their neural-network design and how they are trained, AIs are not good at responding in open-ended situations that require taking context into account. This observation led me to suggest a number of key competencies that we need in order to flourish in the Age of AI. In most of these cases, we can use the CAL to help us deploy the competency effectively and efficiently. We can divide these competencies into two broad categories, Foundational Skills and Essential Knowledge.

Foundational Skills

We all should be able to harness the following skills well, both with the CAL and on our own, not only in order to interact with AIs but also to interact with other humans.

Critical Thinking. As discussed in Chapter 3, critical thinking is not "one thing," and we need to rely on different types of critical thinking in a wide variety of contexts. To deeply grasp the methods in Table 3.1, we need to think statistically, which requires some basic math. Such learning should focus on the concepts, so that we can develop intuitions that help us navigate increasingly complex environments.

Creative Problem Solving. As discussed in Chapter 4, many types of problems require creative approaches, and often require "out of the box" thinking in open-ended situations where context must be taken into account. We need to employ a variety of heuristics for both the divergent thinking phase and the convergent thinking phase of creative problem solving. To engage in either critical thinking or creative problem solving well, we need to know about our human cognitive strengths and limitations, and how to manage the latter (Chapter 5).

Emotional Intelligence. We can develop our capacity not only to recognize and manage our own emotions (Chapter 6), but also to grasp the effects of our actions on other people. In addition, we need to understand other people's emotions and how they are affecting their behavior. The way others react depends in part on their personalities (Chapter 7), which emotional intelligence can help us interpret and respond to (Chapter 8). This skill, along with good communication skills, can go a long way toward facilitating interpersonal harmony.

Human Communication. Another crucial skill, discussed in Chapter 9, is the ability to communicate clearly and appropriately. We need to know how best to communicate different sorts of information, to particular audiences, for specific purposes. As noted earlier, one crucial communication skill for the Age of AI is knowing how to debunk disinformation and misinformation. Communication skills, combined with critical thinking, creative problem solving, and emotional intelligence, are key to ensuring that truth and transparency prevail—and helping society as a whole to adapt to a world infused with AI.

Leading, Following, Collaborating. In Chapter 10 we considered the skills that underlie how we humans work collectively, as leaders, followers, and collaborators. These skills will be increasingly important in the Age of AI, as we humans focus increasingly on our interactions with other humans.

Adapting and Learning. Given how quickly the world is changing, and is likely to keep changing as technology advances, we humans are going to need to adapt and learn over the course of our lives. People need to learn how to do this efficiently and effectively (Chapter 10).

We can use the CAL to help us exercise each of these skills, and/or to help us learn them. Indeed, in many situations, using the CAL to assist us will—as a byproduct—end up helping us to master the skills.

Essential Knowledge

If we humans are to capitalize on what we can do better than current AIs, we are going to need to learn how to assess and adjust to the context in which events unfold. To do this well, we need to learn about a number of domains. Moreover, we will need to keep learning for the rest of our lives because there will always be new relevant discoveries and developments. The domains we need to address include:

Humanities. As discussed earlier in the present chapter, the humanities can illuminate our lives in many ways. They can help us identify and evaluate possible life paths, learn via vicarious experiences, make sense of the world around us, and open up new vistas to consider.

Human Personality. By learning about advances in our understanding of the nature of personality, we not only gain self-knowledge, but also position ourselves to grasp how to best interact with others (Chapter 7).

Social Sciences. The social sciences offer a wealth of useful knowledge. I've illustrated this throughout this book by drawing on various facets of psychology, such as what it tells us about the nature of learning, managing emotion, and characterizing personality and emotional intelligence. In addition, economics can help us decide how to prioritize the ways we allocate resources, such as time and money. By studying economics, we can gain insights into the financial and material consequences of particular human and professional goals. Similarly, sociology leads us to consider social factors that can influence our choices as well as the impacts of our actions on society. In addition, anthropology leads us to adopt a culturally sensitive perspective on the world. Adopting this perspective can help us recognize and address our own cultural blinders, helping us to see possibilities that were not previously evident. In general, theories, concepts, and findings in the social sciences can help us conceive of and evaluate possible human and professional goals.

STEM Fields. We will be led to learn many aspects of Science, Technology, Engineering, and Math (STEM) material when we master critical thinking and creative problem solving. For example, when we consider human judgment and decision making, we need to learn essentials about the brain. And, of course, to stay abreast of developments in AI, we need to be conversant with aspects of computer science. But more than this, we need a general awareness of STEM because the progress in these fields will rock our world. This progress is already stunning and AIs will only accelerate it.[8,9] For example, AIs will accelerate finding cures for diseases, extending the human lifespan, creating new building materials, and providing cheap energy. We need to grasp the fundamentals of STEM fields if only so that we don't feel buffeted by seemingly magical forces outside human control.

Navigating Everyday Contexts. Finally, consider the everyday contexts in which we all function. To the extent that we understand the fundamentals of them, we are better positioned to be able to reason about them when faced with an open-ended situation. Specifically, we should know something about: 1) the legal system, from mundane facts about rights when stopped by a police officer, to basic facts about what lawyers do, to the way the court system operates; 2) the medical system, from what different types of healthcare providers do, to knowing how to interact with them, to understanding medical reports; 3) the financial system, ranging from personal finances, to taxes, to basics of the stock market; 4) everyday physical infrastructure, such as home plumbing systems, home maintenance, home energy use, and how to find and evaluate repair people.

Learning such material gives us a wealth of knowledge about important aspects of our daily lives, which in turn should make us more comfortable in our environments and give us more control over our lives. In every case, AIs are going to have a serious impact, and we need to anticipate how this is going to change our lives and adapt to these changes.

In this book I've stressed two main themes. First, we humans should focus on tasks and situations that require responding to open-ended situations where context must be taken into account. Contemporary AIs are not very good in such situations, and we have reason to believe that they will not overcome these limitations in the near future. Second, AIs can help us carry out such tasks even better than we do now, acting as cognitive amplifiers. We should learn how to interact with AIs to get the most of them, as we forge ahead in this emerging world. AIs are a fact of life, and we should learn to take advantage of them to further our goals, to help us flourish in this new Age of AI.

Digging Deeper

We again end with suggestions of videos that provide accessible, more in-depth treatment of material in this chapter. We also provide search terms to help locate new videos on the topic.

Fareed Zakaria—The Importance of the Liberal Arts: https://www.youtube.com/watch?v=g-Kbfu_XOi0

How to know your life purpose in 5 minutes | Adam Leipzig | TEDxMalibu: https://www.youtube.com/watch?v=vVsXO9brK7M

- "Curating life goals in the Age of AI"
- "Role of liberal arts education in personal development"
- "Humanities and self-understanding"
- "Critical thinking and creative problem solving in personal growth"
- "Using literature and art to reflect on life experiences"
- "Philosophy and ethical foundations of life goals"
- "Epistemology and evaluating knowledge claims"
- "Data, information, knowledge, and wisdom hierarchy"
- "AI-assisted life coaching and goal setting"

The Rise of the Machines—Why Automation is Different this Time: https://www.youtube.com/watch?v=WSKi8HfcxEk

100M jobs gone because of AI: What will we do? https://www.youtube.com/watch?v=xbNDNEBuTgs

AI And The Future Of Jobs: https://www.youtube.com/watch?v=h1PXTqdS9T0

- "Impact of AI on job market"
- "Jobs most likely to be replaced by AI"
- "Jobs that will require working with AI"
- "Jobs least likely to be affected by AI"
- "Essential skills for management roles"
- "Leadership skills training"
- "Strategic thinking in business"

- "Effective communication for managers"
- "Networking strategies for career development"
- "Leveraging social media for professional networking"
- "Mentorship and career growth"

References

1   Santayana, G. (1905). *The life of reason*. Prometheus Books.
2   Harper, T. A. (2024, January 26). The 100 year extinction panic is back, right on schedule. *New York Times*. www.nytimes.com/2024/01/26/opinion/pol ycrisis-doom-extinction-humanity.html?smid=nytcore-ios-share&referrin gSource=articleShare
3   Futuyma, D. J. (2013). *Evolution* (3rd ed.). Sinauer Associates/Oxford University Press.
4   Griffiths, A. J. F., Wessler, S. R., Carroll, S. B., & Doebley, J. (2015). *Introduction to genetic analysis* (11th ed.). W. H. Freeman and Company.
5   Plato. (B. Jowett, Trans.). (2014) *Apology*. CreateSpace Independent Publishing Platform.
6   Ackoff, R. L. (1989). From data to wisdom. *Journal of Applied Systems Analysis, 16*, 3–9.
7   Whitehead, A. N. (1929). *The aims of education and other essays*. Macmillan.
8   Frueh, S. (2023, November 6). How AI is shaping scientific discovery. *National Academies*. www.nationalacademies.org/news/2023/11/how-ai-is-shaping-scientific-discovery
9   Steimer, S. (2023, July 14). UChicago study explores how AI can predict discoveries and who will make them. *UChicago News*. https://news.uchicago.edu/story/human-aware-ai-helps-accelerate-scientific-discoveries-new-resea rch-shows

Index